MATHEMATICS FOR THE MACHINE TRADES

DAVID L. GOETSCH

DEBORAH M. GOETSCH

RAYMOND L. RICKMAN

PRENTICE HALL, Englewood Cliffs, NJ 07632

Library of Congress Cataloging-in-Publication Data

Goetsch, David L.
Mathematics for the machine trades.

1. Shop mathematics. I. Goetsch, Deborah M.
II. Rickman, Raymond L. III. Title.
TJ1165.G58 1988 513'.14'024621 87-13147
ISBN 0-13-563008-8

Editorial/production supervision and
interior design: *Ellen Denning*
Manufacturing buyer: *Lorraine Fumoso*
Cover photograph courtesy of Auatrol, Inc.

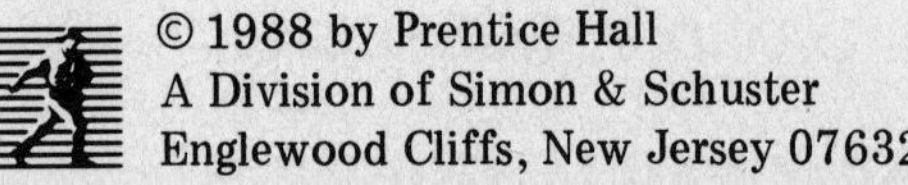

Printed in the United States of America

10 9 8 7 6 5 4 3 2 1

ISBN 0-13-563008-8 025

PRENTICE-HALL INTERNATIONAL (UK) LIMITED, *London*
PRENTICE-HALL OF AUSTRALIA PTY. LIMITED, *Sydney*
PRENTICE-HALL CANADA INC., *Toronto*
PRENTICE-HALL HISPANOAMERICANA, S.A., *Mexico*
PRENTICE-HALL OF INDIA PRIVATE LIMITED, *New Delhi*
PRENTICE-HALL OF JAPAN, INC., *Tokyo*
SIMON & SCHUSTER ASIA PTE. LTD., *Singapore*
EDITORA PRENTICE-HALL DO BRASIL, LTDA., *Rio de Janeiro*

Dedicated to our little angel, Savannah Day, with love . . .

CONTENTS

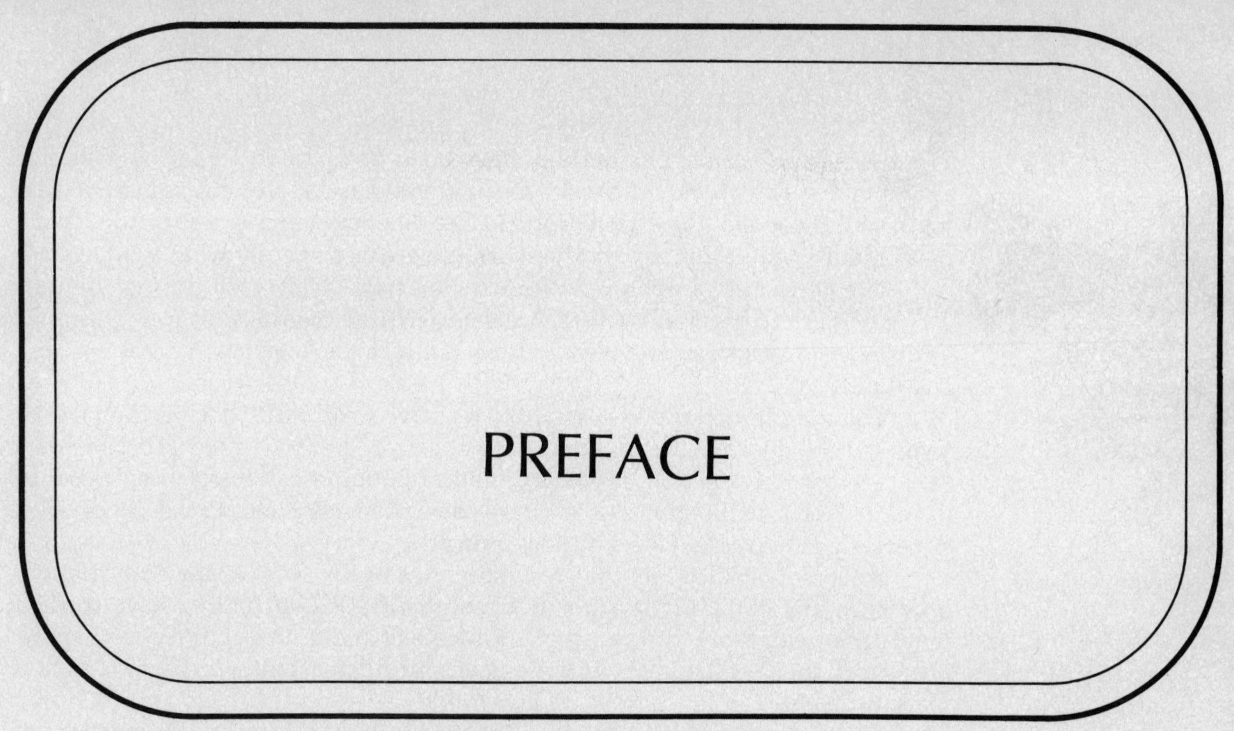

PREFACE

WHO IS THIS BOOK FOR?

This book is for three groups of people: (1) students who wish to become machine technicians, (2) practicing machining technicians, and (3) persons working in other occupations who would like or who need to make a career change. *Mathematics for the Machine Trades* is designed for use in both a traditional classroom setting and in nontraditional settings such as workshops, seminars, or on-the-job learning situations. Any person in any setting who needs to learn the fundamentals of math as well as how to apply those fundamentals in a "real world" setting will find this book helpful.

WHAT TYPES OF SCHOOLS AND AGENCIES MIGHT USE THIS BOOK?

Any institution or agency providing technical training will find *Mathematics for the Machine Trades* to be appropriate for use. Such institutions and agencies include community colleges, technical colleges, vocational and technical schools, private educational agencies, government training agencies, military training agencies, and private-sector companies that offer on-the-job training.

PREREQUISITES

This book has been especially structured to require no prerequisites. Part 1 is a special, self-paced arithmetic review for students who graduated from high school without a solid grasp of arithmetic or for practitioners who have not used arithmetic for a prolonged period of time. It may be used as a separate course that is part of a training program, as an integrated part of a self-paced training program, or as a learning tool in workshops, seminars, or on-the-job training situations.

NEED FOR THE BOOK

For the past 10 years, the authors have been involved in preparing technical students for the world of work. During this time it became apparent that math skills needed close attention. It also became apparent that the "technical math" books already on the market were not exactly what was needed.

The consensus among practitioners we have dealt with during the past 10 years is that the hands-on technician needs to have a solid foundation of knowledge and skills in three critical math areas: algebra, geometry, and trigonometry.

The consensus among educators we have dealt with during the past 10 years is that many students come into their programs ill-prepared to undertake a study of algebra, geometry, and trigonometry. First, they need to develop basic arithmetic skills. Numerous nationally circulated reports on America's public schools echo their thoughts.

Practitioners also feel that too many beginning technicians do not know how to apply their math skills in a real-world setting. If they have studied math, they can work the types of problems found at the end of the chapter, but seem to have trouble working math problems that are part of actual work projects.

Educators also feel that most technical math books on the market try to cover too much territory. They attempt to cover, in one book, arithmetic through calculus. The result is that the book overpowers or frightens students. Additionally, most technical math books attempt to include problems for several different technical programs such as drafting, electronics, automotive, heating, ventilating, and cooling, machining, and so forth. In so doing, they lose the specificity that promotes understanding, acceptance on the part of students, and learning.

Mathematics for the Machine Trades was written in response to the stated needs of both practitioners and educators. It was especially designed to include those characteristics they felt were positive and to exclude those they felt were negative.

INNOVATIONS

This book, because it was designed to meet specific needs, has a number of important built-in innovations that set it apart from other books:

1. The format is structured to allow the book to be used in a traditional classroom, a self-paced classroom, a workshop setting, or an on-the-job training situation.
2. A special three-part teaching and learning approach is utilized in which students first learn or review basic arithmetic; then develop the essential skills from the critical math areas of algebra, geometry, and trigonometry without the additional burden of trying to apply those skills to actual real world projects; and finally, once all the necessary skills have been developed, apply them to a wide range of actual work-world projects that include math problems which mirror those that will be confronted on the job.
3. The book is divided into three distinct parts, each especially designed to meet specific needs. Consequently, each part has its own format and special features.

With minor adaptions, portions of Parts 1 and 2 and the Appendices are taken from *Math for Computer-Aided Drafting* by David L. Goetsch and Deborah M. Goetsch, Prentice-Hall, Inc., © 1987. This material is adapted by permission of Prentice-Hall, Inc.

MAJOR FEATURES

The major features of the book must be presented in three different parts which match the three parts of the book. The major features of Part 1 include:

1. Pretests and posttests for each chapter so that students who already have the skills covered can skip that which is known if the instructor wishes them to.
2. Practical application problems within each chapter for each new concept presented. This allows for immediate application of the concept that is being learned.
3. A comprehensive set of review problems at the end of each chapter that provides learners with opportunities to apply their learning for all of the concepts presented in a given chapter.

The major features of Part 2 include:

1. Practical application problems within each chapter for each new concept presented. This allows for immediate application of the concept that is being learned.
2. A comprehensive set of review problems at the end of each chapter that provides learners with opportunities to apply their learning for all of the concepts presented in a given chapter.

The major features of Part 3 include:

1. Each chapter is concerned with nothing but application and contains real-world projects.
2. The projects are to be accomplished as if in an actual work setting. To do so, learners will have to apply all of the skills learned in Parts 1 and 2.

Other major features of the book include:

1. An extensive appendix for ready reference in solving problems.
2. Answers to all pretest and posttest problems from Part 1.
3. Answers to the odd-numbered practical application problems for Parts 1 and 2.
4. Answers to the odd-numbered review problems for Parts 1 and 2.

DAVID L. GOETSCH
DEBORAH M. GOETSCH
RAYMOND L. RICKMAN

INTRODUCTION

The machine trades are among the most important of the various manufacturing occupations. Almost every manufactured product contains parts that had to be machined. Such traditional operations as turning, cutting, boring, and shaping are all still fundamental processes in the modern manufacturing shop. However, how these processes are accomplished is changing. Traditionally, these processes have been accomplished mechanically or by hand. Advances in computer technology and other forms of automation are changing this.

The machining trades are in a state of transition from mechanical and manual operations to computer-controlled automated processes. Some of the more frequently heard terms in the language of machining are computer-aided manufacturing (CAM), computer numerical control (CNC), computer integrated manufacturing (CIM), robotics, and flexible manufacturing systems (FMS).

This transition from mechanical to automated processes is having an impact on the knowledge and skills needed by practitioners of the various machine trades. By eliminating much of the repetitive hands-on work required to move a product through production, automation is giving machining personnel more time for mind work. This is a positive step because it allows them to spend more time finding ways to improve both the product and productivity in producing it. However, it has also made cognitive capabilities such as math skills even more important than ever for machine trades personnel.

Before beginning a study of math for the machine trades, students should know the types of math skills they will need on the job for today and tomorrow and some of the more common ways these skills will be applied in the modern machine shop.

MATH SKILLS NEEDED IN THE MODERN MACHINE SHOP

There are several categories or levels of math. The categories that are most important to students in the machine trades are arithmetic, algebra, geometry, and trigonometry. Within each of these categories there are a number of specific skill areas that are important. Those that represent the minimum skills required of the modern machinist are listed in the subsections that follow.

Minimum Arithmetic Skills Needed by Machining Personnel

At a minimum, students in the machine trades should be able to do the following:

1. Add, subtract, multiply, and divide whole numbers
2. Add, subtract, multiply, and divide common fractions
3. Add, subtract, multiply, and divide mixed numbers
4. Add, subtract, multiply, and divide decimal fractions
5. Convert from common fractions to decimal fractions
6. Convert from decimal fractions to common fractions
7. Raise numbers to powers
8. Take roots of numbers
9. Calculate percentages

Minimum Algebra Skills Needed by Machining Personnel

At a minimum, students in the machine trades should be able to do the following:

1. Apply symbolism in solving problems
2. Manipulate signed numbers
3. Perform basic algebraic operations
4. Solve equations
5. Apply standard shop formulas

Minimum Geometry Skills Needed by Machining Personnel

At a minimum, students in the machine trades should be able to do the following:

1. Solve problems involving the geometric principles of lines
2. Solve problems involving the geometric principles of angles
3. Solve problems involving the geometric principles of triangles
4. Solve problems involving the geometric principles of polygons
5. Solve problems involving the geometric principles of circles
6. Solve problems involving the geometric principles of arcs

Minimum Trigonometry Skills Needed by Machining Personnel

At a minimum, students in the machine trades should be able to do the following:

1. Solve right triangles
2. Solve oblique triangles
3. Calculate the area of triangles

APPLYING MATH SKILLS IN THE MODERN MACHINE TRADES

It is important to develop the math skills listed in the preceding section, but just having the skills is not enough. You must also be able to apply these skills in a work setting. There are many ways that math skills are applied on the job. Some of the most frequent applications in the machine trades are in the following:

1. Calculating dimensions
2. Taking inventory
3. Calculating cost
4. Calculating weight, area, and volume
5. Calculating time and wages
6. Performing thread calculations
7. Determining the amount of waste
8. Using shop formulas
9. Solving equations
10. Calculating thread values
11. Performing gear calculations
12. Performing speed and feed calculations
13. Calculating turning revolutions per minute
14. Performing taper calculations
15. Calculating gear ratios
16. Calculating fluid mix ratios
17. Performing output calculations
18. Performing pulley calculations
19. Performing angle calculations
20. Solving right triangles
21. Solving oblique triangles
22. Performing sine bar calculations
23. Performing bevel calculations
24. Calculating missing dimensions
25. Calculating cutting depths
26. Performing dovetail calculations

The chapters that follow are especially designed to help you develop the fundamental math skills needed in the machine trades and how to apply them on the job.

PART 1
Arithmetic Review

1
WHOLE NUMBERS

INTRODUCTION TO PART 1

Part I is a comprehensive review of arithmetic. It consists of five chapters and has been especially designed to serve several purposes: (1) a comprehensive review for learners who have already developed arithmetic skills but need to review them before pursuing a study of algebra, geometry, and trigonometry; (2) an opportunity to develop arithmetic skills for those learners who either graduated from or left high school without them; and (3) a quick reference for learners who might need to refer back to certain arithmetic concepts on occasion.

Each chapter has been especially structured to allow learners to skip over material they already know if they do not wish to review it. The Pretest* for each chapter in Part 1 gives learners an opportunity to measure their knowledge of the material covered in that chapter. If a learner scores 90% or better on a Pretest, the chapter can be skipped (with the instructor's permission). A score of less than 90% indicates that a review of the material covered in that chapter is in order.

Each chapter in Part 1 also contains a Posttest. All learners studying a given chapter in Part 1 should complete the Posttest for that chapter before proceeding to the next chapter. A socre of less than 90% on the Posttest indicates that additional work is in order before proceeding. Learners may find it helpful to work the Practical Application problems within each chapter and the Review Problems at the end of each chapter before attempting the Posttest.

Learners who are able to work the problems set forth in each chapter and at the end of each chapter are then ready to proceed to Part 2 for a study of the essentials of algebra, geometry, and trigonometry. Chapter 1 covers whole numbers and the various mathematical operations that technicians need to be able to perform on whole numbers. Chapter 2 covers common fractions and the operations technicians need to be able to perform using common fractions. Chapter 3 covers mixed numbers and operations with

*Pretests and Posttests for Chapters 1 through 5 appear in Appendix B.

mixed numbers. Chapter 4 deals with decimal fractions and operations with decimal fractions. Chapter 5 concludes Part 1 with a discussion of powers, roots, percentages, and associated operations.

Technicians must be proficient in mathematics. Such proficiency begins with an understanding of whole numbers. Whole numbers are those we know as the *counting numbers* (0, 1, 2, 3, 4, 5, 6, 7, 8, 9, . . .). Whole numbers form the basis of all other numbers used in mathematics. In fact, all other numbers can be defined in terms of whole numbers.

To assist you in recalling your understanding of whole numbers, this chapter covers the following major concepts: expressing place values of whole numbers, expanding whole numbers, reading and writing whole numbers, adding whole numbers, subtracting whole numbers, multiplying whole numbers, and dividing whole numbers.

EXPANDING PLACE VALUES OF WHOLE NUMBERS

In the decimal system, all numbers can be formed using 10 digits (0, 1, 2, 3, 4, 5, 6, 7, 8, and 9). Each digit represents a certain value depending on its location or "place" in the number. This value is known as the *place value* of the digit. Place values begin with units and go up infinitely. The place values for whole numbers representing billions or less are: units, tens, hundreds, thousands, ten thousands, hundred thousands, millions, ten millions, hundred millions, and billions.

The chart in Figure 1-1 shows how place values in numbers can be determined. To determine the place values of digits within a number, begin with the right-hand-most digit in the number. This digit represents units or ones. The next digit to the left represents tens. The next digit to the left represents hundreds. Notice that each digit in a number represents a place value 10 times the value of the digit to its immediate right. In Figure 1-1 the place value of the 5 is billions, of the 6 is millions, of the 8 is thousands, and of the 3 is units or ones.

5	BILLIONS
4	HUNDRED MILLIONS
7	TEN MILLIONS
6	MILLIONS
2	HUNDRED THOUSANDS
0	TEN THOUSANDS
8	THOUSANDS
9	HUNDREDS
1	TENS
3	UNITS

Figure 1-1

SAMPLE PROBLEM AND SOLUTION

Express the place value of the underlined digit in the following number:

572,852 (hundreds)

SAMPLE PROBLEM AND SOLUTION

Express the place value of the underlined digit in the following number:

678,561,211 (hundred-thousands)

SAMPLE PROBLEM AND SOLUTION

Express the place value of the underlined digit in the following number:

1,051 (thousands)

PRACTICE PROBLEM SET 1-1

Express the place value of the digit underlined in each number.

1. 16,847	2. 32	3. 179	4. 156,305
5. 1,712,869	6. 5,082	7. 10	8. 15,973
9. 56,823,970	10. 3,564		

EXPANDING WHOLE NUMBERS

We use the digits 2 and 8, written as 28, to represent the number twenty-eight. This (28) is actually a short way of writing 2 tens plus 8 units or ones. Whole numbers can be expanded by using a three-step procedure:

1. Determining the place value of each digit
2. Writing the digit times its place value in parentheses
3. Separating each set of parentheses with a plus sign (+)

Using this procedure, the number 762 can be expanded as follows:

$$(7 \times 100) + (6 \times 10) + (2 \times 1)$$

SAMPLE PROBLEM AND SOLUTION

Expand the number 981.

$$(9 \times 100) + (8 \times 10) + (1 \times 1)$$

SAMPLE PROBLEM AND SOLUTION

Expand the number 6,854.

$$(6 \times 1{,}000) + (8 \times 100) + (5 \times 10) + (4 \times 1)$$

SAMPLE PROBLEM AND SOLUTION

Expand the number 79,845.

$$(7 \times 10{,}000) + (9 \times 1{,}000) + (8 \times 100) + (4 \times 10) + (5 \times 1)$$

PRACTICE PROBLEM SET 1-2

Expand each whole number.

1. 172	2. 3	3. 16	4. 2,283	5. 99
6. 54,333	7. 7,524	8. 4	9. 62	10. 413

READING AND WRITING WHOLE NUMBERS

Once you know how to express the place values of whole numbers and to expand whole numbers, they may be easily read and written. Notice in the preceding sections that whole numbers with more than three digits were grouped using commas (3,266,759). This is the first step in reading or writing whole numbers. Whole numbers are read and written as word statements. For example, the number 1,617 is written one thousand, six hundred seventeen.

Before attempting to read or write a whole number containing more than three digits, group the digits using commas. To do this, begin with the right-hand-most digit and work your way to the left making groups of three digits each, as in the following example:

4,629,812

To read or write this number as a word statement, work from left to right using the words that represent each number according to its grouping. The number in the example above is read or written: four million, six hundred twenty-nine thousand, eight hundred twelve. Notice that the commas in the word statement match the commas in the number.

SAMPLE PROBLEM AND SOLUTION

Write the word statement for 819.

eight hundred nineteen

SAMPLE PROBLEM AND SOLUTION

Write the word statement for 4,132.

four thousand, one hundred thirty-two

PRACTICE PROBLEM SET 1-3

Write each whole number as a word statement.

1. 719	2. 64	3. 10,419	4. 215,333
5. 7,326,424	6. 201	7. 8,888	8. 97,615
9. 112,537	10. 5,233,648		

ADDING WHOLE NUMBERS

Technicians must be able to add whole numbers. The addition of whole numbers involves three elements: the addends, the plus sign, and the sum.

The *addends* are the numbers to be added. The *plus sign* (+) indicates that the process of addition is to be performed. The *sum* is the result of the process. These three elements are illustrated in the following example:

	123	addend
plus sign	+ 411	addend
	534	sum

To add whole numbers, you must know the complete set of sums of all pairs of single-digit numbers. These sums are illustrated in the chart in Figure 1-2.

Whole numbers are prepared for addition by stacking them in columns. Make sure that digits in the units, tens, hundreds, and so on, places line up, as in the following example:

	hundreds	tens	units
	7	6	2
	5	1	3
+	4	5	5

+	0	1	2	3	4	5	6	7	8	9
0	0	1	2	3	4	5	6	7	8	9
1	1	2	3	4	5	6	7	8	9	10
2	2	3	4	5	6	7	8	9	10	11
3	3	4	5	6	7	8	9	10	11	12
4	4	5	6	7	8	9	10	11	12	13
5	5	6	7	8	9	10	11	12	13	14
6	6	7	8	9	10	11	12	13	14	15
7	7	8	9	10	11	12	13	14	15	16
8	8	9	10	11	12	13	14	15	16	17
9	9	10	11	12	13	14	15	16	17	18

Figure 1-2

Once the numbers have been prepared for addition, the addition can be performed. The addition operation for the example above would be performed as follows:

$$\begin{array}{r} 7\ 6\ 2 \\ 5\ 1\ 3 \\ +\ \underline{4\ 5\ 5} \\ 1\ ,\ 7\ 3\ 0 \end{array}$$

Step 1

Add the numbers in the units column from top to bottom.

$$2 + 3 + 5 = 10$$

Write the 0 under the units column and add the 1 to the top of the tens column.

Step 2

Add the numbers in the tens column from top to bottom.

$$(1) + 6 + 1 + 5 = 13$$

Write the 3 under the tens column and the 1 to the top of the hundreds column.

Step 3

Add the numbers in the hundreds column from top to bottom.

$$(1) + 7 + 5 + 4 = 17$$

Write the 7 under the hundreds column and the 1 to the left of it. Separate the 1 and the 7 with a comma.

SAMPLE PROBLEMS AND SOLUTIONS

Complete each addition problem.

$$\begin{array}{r} 56 \\ +\ \underline{12} \\ 68 \end{array} \qquad \begin{array}{r} 17 \\ +\ \underline{88} \\ 105 \end{array} \qquad \begin{array}{r} 694 \\ +\ \underline{\ \ 57} \\ 751 \end{array} \qquad \begin{array}{r} 4{,}682 \\ +\ \underline{9{,}479} \\ 14{,}161 \end{array}$$

PRACTICE PROBLEM SET 1-4

Complete each addition problem.

1. $\begin{array}{r} 54 \\ +\ \underline{37} \end{array}$

2. $\begin{array}{r} 85 \\ +\ \underline{15} \end{array}$

3. $\begin{array}{r} 79 \\ +\ \underline{64} \end{array}$

4. $\begin{array}{r} 26 \\ +\ \underline{33} \end{array}$

5. $\begin{array}{r} 116 \\ +\ \underline{732} \end{array}$

6. $\begin{array}{r} 432 \\ +\ \underline{444} \end{array}$

7. $\begin{array}{r} 839 \\ +\ \underline{567} \end{array}$

8. $\begin{array}{r} 791 \\ +\ \underline{232} \end{array}$

9. $\begin{array}{r} 10{,}515 \\ 4{,}617 \\ +\ \underline{19{,}821} \end{array}$

10. $\begin{array}{r} 112{,}469 \\ 456{,}789 \\ +\ \underline{321{,}432} \end{array}$

SUBTRACTING WHOLE NUMBERS

Technicians spend a great deal of time subtracting whole numbers. Subtraction is the inverse of addition and is used hand in hand with addition.

The subtraction of whole numbers involves four elements: the subtrahend, the minuend, the minus sign, and the remainder. The *subtrahend* is the number to be subtracted. The *minuend* is the number from which the subtrahend is subtracted. The *minus sign* (−) indicates that the process of subtraction is to be performed. The *remainder* is the result of the subtraction process. These four elements are illustrated in the following example:

		146	minuend
minus sign	−	123	subtrahend
		23	remainder

Whole numbers are prepared for subtraction by placing the subtrahend under the minuend so that individual digits form columns according to place values. Make sure that the units, tens, hundreds, thousands, and so on, digits in the subtrahend and minuend line up, as in the following example:

	thousands		hundreds	tens	units
	1	,	4	8	7
−			2	8	3

Once the numbers have been prepared for subtraction, the subtraction operation can be performed. The subtraction operation for the example above would be performed as follows:

Step 1

Begin in the units column. Subtract the 3 in the subtrahend from the 7 in the minuend. From the addition table learned in the preceding section, you know that the remainder is 4.

$$\begin{array}{r} 1,487 \\ -\quad 283 \\ \hline 4 \end{array}$$

Step 2

Move to the tens column and subtract the 8 in the subtrahend from the 8 in the minuend. From the addition table learned in the preceding section, you know that the remainder is 0.

$$\begin{array}{r} 1,487 \\ -\quad 283 \\ \hline 04 \end{array}$$

Step 3

Move to the hundreds column and subtract the 2 in the subtrahend from the 4 in the minuend. From the addition table learned in the preceding section, you know that the remainder is 2.

$$\begin{array}{r} 1,487 \\ -\quad 283 \\ \hline 204 \end{array}$$

Step 4

Move to the thousands column. Since there is no digit in the subtrahend, simply bring down the 1 in the minuend to complete the remainder. The remainder is 1,204.

$$\begin{array}{r} 1,487 \\ -\quad 283 \\ \hline 1,204 \end{array}$$

You may check the remainder by adding it to the subtrahend. The result should equal the minuend, as in the following example:

$$\begin{array}{r} 1,204 \\ +\quad 283 \\ \hline 1,487 \end{array}$$

In the subtraction example just presented, each digit in the minuend had a value greater than its corresponding digit in the subtrahend. This simplifies the subtraction process. However, you will not always be this lucky. Frequently, a digit in the minuend will have a value that is less than its corresponding digit in the subtrahend. When this is the case, an extra step is required in the subtraction process. That step is known as *borrowing.*

Borrowing involves increasing the value of a digit in the minuend by 10 by borrowing 10 units from the digit to its immediate left. You are able to do this because, as you learned in expanding whole numbers, each digit is a whole number and has a place value 10 times greater than the digit to its immediate right.

Borrowing increases the value of the number being borrowed from by 10 and decreases the value of the number borrowed from by 1. The subtraction process with borrowing is accomplished as follows:

$$\begin{array}{r} 8,575 \\ -\quad 466 \\ \hline \end{array}$$

A digit in the subtrahend is greater than its corresponding digit in the minuend

Step 1

Begin in the units column. You will notice immediately that the 6 in the subtrahend is greater than the 5 in the minuend, which indicates that you will have to borrow. Increase the 5 to 15 by borrowing from the 7 in the tens column. This will decrease the 7 to 6.

$$\begin{array}{r} 6 \\ 8,5\not{7}\,^{1}5 \\ -\quad 466 \\ \hline \end{array}$$

Step 2

Complete the subtraction process using the normal steps just learned to compute the remainder.

$$\begin{array}{r} 8,575 \\ -\quad 466 \\ \hline 8,109 \end{array}$$

Step 3

Check your work by adding the remainder to the subtrahend.

$$\begin{array}{r} 8{,}109 \\ +\quad 466 \\ \hline 8{,}575 \end{array}$$

PRACTICE PROBLEM SET 1-5

Complete each subtraction problem.

1. $\begin{array}{r} 77 \\ -\ 32 \\ \hline \end{array}$
2. $\begin{array}{r} 179 \\ -\ 58 \\ \hline \end{array}$
3. $\begin{array}{r} 1{,}432 \\ -\ 1{,}212 \\ \hline \end{array}$
4. $\begin{array}{r} 12{,}788 \\ -\ 6{,}511 \\ \hline \end{array}$
5. $\begin{array}{r} 453 \\ -\ 78 \\ \hline \end{array}$
6. $\begin{array}{r} 8{,}888 \\ -\ 7{,}999 \\ \hline \end{array}$
7. $\begin{array}{r} 12{,}481 \\ -\ 3{,}764 \\ \hline \end{array}$
8. $\begin{array}{r} 150{,}719 \\ -\ 49{,}864 \\ \hline \end{array}$
9. $\begin{array}{r} 46{,}324 \\ -\ 45{,}666 \\ \hline \end{array}$
10. $\begin{array}{r} 56{,}471 \\ -\ 6{,}112 \\ \hline \end{array}$

MULTIPLYING WHOLE NUMBERS

Multiplication of whole numbers is an important concept for technicians. It is used in many ways, such as computing area, volume, stress, and loads, or in making numerous other computations. Multiplication is really just a way to shortcut the addition process. For example, examine the following addition process:

$$\begin{array}{r} 12 \\ 12 \\ 12 \\ +\ 12 \\ \hline 48 \end{array}$$

Using multiplication instead of addition allows the problem to be simplified to

$$\begin{array}{r} 12 \\ \times\ \ 4 \\ \hline 48 \end{array}$$

The multiplication of whole numbers involves four elements: the multiplicand, the multiplier, the product, and the multiplication or times sign. The *multiplicand* is the number to be multiplied. The *multiplier* is the number by which the multiplicand is multiplied. The *product* is the result of the multiplication process. The *times sign* (X) indicates that the multiplication process is to be performed. These four elements are illustrated in the following example:

$$\begin{array}{rl} 172 & \text{multiplicand} \\ \times\quad 15 & \text{multiplier} \\ \hline 2{,}580 & \text{product} \end{array}$$

X	1	2	3	4	5	6	7	8	9	10	11	12
1	1	2	3	4	5	6	7	8	9	10	11	12
2	2	4	6	8	10	12	14	16	18	20	22	24
3	3	6	9	12	15	18	21	24	27	30	33	36
4	4	8	12	16	20	24	28	32	36	40	44	48
5	5	10	15	20	25	30	35	40	45	50	55	60
6	6	12	18	24	30	36	42	48	54	60	66	72
7	7	14	21	28	35	42	49	56	63	70	77	84
8	8	16	24	32	40	48	56	64	72	80	88	96
9	9	18	27	36	45	54	63	72	81	90	99	108
10	10	20	30	40	50	60	70	80	90	100	110	120
11	11	22	33	44	55	66	77	88	99	110	121	132
12	12	24	36	48	60	72	84	96	108	120	132	144

Figure 1-3

Hand-held calculators and computers have simplified the multiplication process considerably. However, before using calculators or computers, it is important that technicians understand how both short and long multiplication are performed manually.

Before attempting to multiply whole numbers, you must learn, at minimum, the products of whole numbers from 1 to 12. These products are presented in Figure 1-3. Take time to memorize this table before attempting to work multiplication problems manually.

Short Multiplication

Short multiplication is the name given to the process when the multiplier is a one-digit number, as in the following example:

$$\begin{array}{r} 737 \\ \times \underline{\quad\ 4} \end{array} \text{ (one-digit multiplier)}$$

This example of a short-multiplication problem would be solved as follows:

Step 1

Arrange the problem so that the multiplier is in the units column under the multiplicand, as in the example above. Multiply the 4 times the 7 that occupies the units place in the multiplicand. The product is 28.

Step 2

Write the 8 under the units column and carry the 2 over to the 10 column.

```
    2
  7 3 7
×     4
      8
```

Step 3

Multiply the 4 times the 3 that occupies the tens place in the multiplicand. The product of this operation is 12. Add the 2 carried over from the units column to the 12 for a sum of 14. Write the 4 under the tens column and carry the 1 over to the hundreds column.

```
  1 2
  7 3 7
×     4
    4 8
```

Step 4

Multiply the 4 times the 7 that occupies the hundreds place in the multiplicand. The product of this operation is 28. Add the 1 carried over from the tens column to the 28 for a sum of 29. Write the 9 under the hundreds column and the 2 immediately to its left.

```
      1 2
      7 3 7
    ×     4
  2 , 9 4 8
```

PRACTICE PROBLEM SET 1-6

Complete each multiplication problem.

1. 19 × 7	2. 22 × 5	3. 334 × 8	4. 415 × 6	5. 4,562 × 9
6. 8,976 × 2	7. 6,482 × 4	8. 1,135 × 8	9. 18,521 × 6	10. 99,960 × 8

Long Multiplication

Long multiplication is the name given to the process when the multiplier contains two or more digits, as in the following example:

```
  7,9 2 3
×   4 6 9   (multiple-digit multipliers)
```

This example of long multiplication would be solved as follows:

Step 1

Arrange the problem so that the units, tens, and hundreds places in the multiplicand and multiplier line up to form columns as in the example above. Multiply the 9 times the 3, then the 2, then the 9, and then the 7 in the multiplicand using the steps outlined for short multiplication.

```
    8 2 2
    7,9 2 3
 ×    4 6 9
 7 1,3 0 7
```

Step 2

Beginning again with the 3 in the units place and working to the left, multiply the 6 in the multiplier times each digit in the multiplicand. Shift your answer so that the right-hand-most digit is written under the tens column, as in the following example:

```
      5 1 1
      7,9 2 3
   ×    4 6 9
    7 1 3 0 7
  4 7 5 3 8
```

Step 3

Beginning again with the 3 in the units place and working to the left, multiply the 4 in the multiplier times each digit in the multiplicand. Shift your answer so that the right-hand-most digit is written under the hundreds column, as in the following example:

```
        3   1
        7,9 2 3
     ×    4 6 9
      7 1 3 0 7
    4 7 5 3 8
  3 1 6 9 2
```

Step 4

Using the normal addition procedures, add the three subproducts shown above to determine the final product of the multiplication process, as in the following example:

```
        7,9 2 3
     ×    4 6 9
      7 1 3 0 7
    4 7 5 3 8
  3 1 6 9 2
  3,7 1 5,8 8 7
```

PRACTICE PROBLEM SET 1-7

Complete each multiplication problem.

1. 75 × 25	**2.** 69 × 43	**3.** 351 × 94	**4.** 888 × 76	**5.** 965 × 846
6. 672 × 561	**7.** 1,477 × 621	**8.** 2,364 × 554	**9.** 11,568 × 477	**10.** 14,999 × 678

DIVIDING WHOLE NUMBERS

Division of whole numbers is an important concept for technicians.

You remember that multiplication is a short way of adding numbers. Division is a short way of subtracting numbers. For example, to determine how many 5's are contained in the number 25, you could use subtraction, as illustrated in the following example:

$25 - 5 = 20$	one step
$20 - 5 = 15$	two steps
$15 - 5 = 10$	three steps
$10 - 5 = 5$	four steps
$5 - 5 = 0$	five steps

Using subtraction it took five steps to determine how many 5's are contained in 25. Using division, these five steps can be reduced to one:

$$25 \div 5 = 5$$

The division process involves five elements: the dividend, the divisor, the quotient, the remainder, and the division sign. The *dividend* is the number that is to be divided. The *divisor* is the number that is divided into the dividend. The *quotient* is the result of the division process. The *remainder* is any number left over after the division process. The *division sign* ($\overline{)}$ or $\div$) indicates that the division process is to be performed. These five elements are illustrated in the following examples:

quotient	7	division sign
divisor	$7\overline{)50}$	dividend
	$-\underline{49}$	
	1	remainder

dividend	division sign	divisor		quotient		remainder
50	$\div$	7	=	7	and	1

Hand-held calculators and computers have simplified the division process considerably. However, before using calculators or computers, it is important that technicians understand how to perform long divisions manually.

To perform long division, you must know the multiplication products presented earlier in Figure 1-3 and rules covering how to deal with the number 0 in dividing whole numbers.

Zero may be used as a dividend. When this is the case, remember the following rule:

Zero divided by any other number equals zero.

$$0 \div 1 = 0$$
$$0 \div 2 = 0$$
$$0 \div 3 = 0$$
$$0 \div 4 = 0$$
$$0 \div (\text{any number}) = 0$$

Zero *may not* be used as a divisor. It is not possible to divide a number by zero.

Long Division

When working a division problem manually, use the long-division symbol $/\overline{\quad}$, as in the following example:

$$8/\overline{415}$$

This example of a long-division problem would be solved as follows:

Step 1

Arrange the problem so that the dividend is housed under the division symbol with the divisor outside the symbol, as in the example above.

Step 2

Begin the division process by attempting to divide the 8 into the first digit of the dividend. Because the divisor (8) is greater than the first digit of the dividend (4), you must move to the second digit of the dividend.

Step 3

Divide the 8 into 41. Eight will go into 41 five times. Write 5 above the division symbol, making sure that it aligns with the 1 in the dividend.

$$\begin{array}{r} 5 \\ 8/\overline{4\ 1\ 5} \end{array}$$

Step 4

Multiply the 5 in the answer times the divisor (8) and place the product (40) under the 41 in the dividend. Subtract 40 from 41 and place the difference under the 40 as shown.

$$\begin{array}{r} 5 \\ 8/\overline{4\ 1\ 5} \\ -\underline{4\ 0} \\ 1 \end{array}$$

Step 5

Attempt to divide the divisor (8) into the remainder (1). Of course, 8 will not go into 1. Therefore, you must bring down the 5 from the dividend.

```
       5
   8/4 1 5
   -4 0
     1 5
```

Step 6

Divide the divisor (8) into 15 and record the answer (1) above the division symbol. Multiply the 1 in your answer times the divisor (8) and record the product (8) under the 15 as shown.

```
       5 1
   8/4 1 5
   -4 0
     1 5
   -   8
```

Step 7

Subtract the 8 from the 15 and record the difference (7). Since there are no additional digits in the dividend, 7 is the remainder. The remainder should be labeled with a capital "R."

```
       5 1
   8/4 1 5
   -4 0
     1 5
   -   8
       7  R
```

The same steps are used when the divisor contains two or more digits. Long-division problems may be checked using multiplication and addition.

To check a long-division problem, multiply the quotient times the divisor. Then add the remainder to the product of this operation as illustrated below. Your final answer should be the dividend.

```
     5 1
   ×   8
   4 0 8
 +     7
   4 1 5  correct
```

PRACTICE PROBLEM SET 1-8

Rewrite each problem using the long-division symbol before performing the division operation.

1. 39 ÷ 6
2. 57 ÷ 7
3. 88 ÷ 3
4. 146 ÷ 9
5. 857 ÷ 4
6. 971 ÷ 14
7. 473 ÷ 17
8. 1,844 ÷ 132
9. 6,662 ÷ 413
10. 10,111 ÷ 584

REVIEW PROBLEMS

Express the place value of the digit underlined in each number.

1. $\underline{8}$,392
2. $\underline{6}$9
3. 1$\underline{0}$1,101

Expand each whole number.

4. 601   5. 2,543

Write each whole number as a word statement.

6. 309   7. 6,905,900

Complete each problem.

8. $69 + 74$

9. $6{,}549 + 605 + 1{,}981$

10. $9{,}844 - 399$

11. $16{,}902 - 10{,}999$

12. 18×6

13. 139×8

14. $6{,}849 \times 3$

15. 999×82

16. 453×345

17. 85×25

18. $98 \div 6$

19. $26\overline{)426}$

20. $100\overline{)377}$

2

COMMON FRACTIONS

Having reviewed whole numbers, you are now ready to move into the next phase of your arithmetic review. This phase deals with a review of fractions. Technicians must be proficient in the use of both common and decimal fractions. This chapter deals with common fractions. Decimal fractions are reviewed in Chapter 4.

To assist you in recalling your understanding of common fractions, this chapter covers the following major concepts: definition of common fractions, expressing fractions as equivalents, reducing fractions to the simplest terms, adding common fractions, subtracting common fractions, multiplying common fractions, and dividing common fractions.

DEFINITION OF COMMON FRACTIONS

A fraction is a given portion of a larger whole. For example, a piece of pie is a fraction of a pie. If you cut a pie into eight pieces and take one, you have taken 1/8 of the pie. Common fractions are used a great deal for expressing dimensions, as in the sketch in Figure 2-1.

Common fractions contain three elements: the *numerator*, the *denominator*, and the *slash* or *bar*. The slash (/) and bar (−) are interchangeable signs either of which may be used when writing common fractions. The elements of a common fraction are illustrated in the following examples:

$$\text{bar}\ \frac{3}{4}\ \begin{matrix}\text{numerator}\\ \text{denominator}\end{matrix}$$

slash

numerator ⟶ 1/2 ⟵ denominator

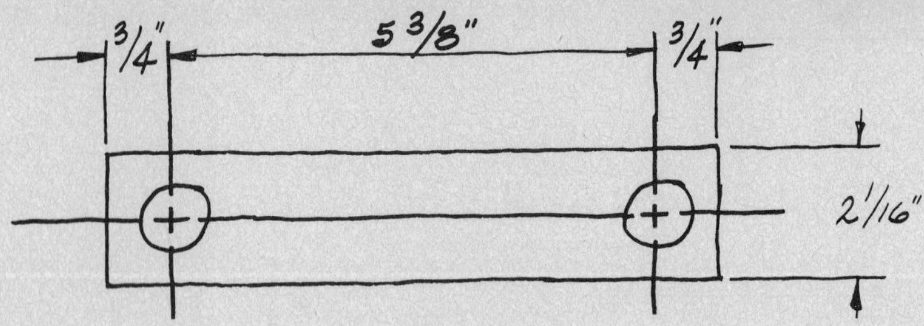

Figure 2-1

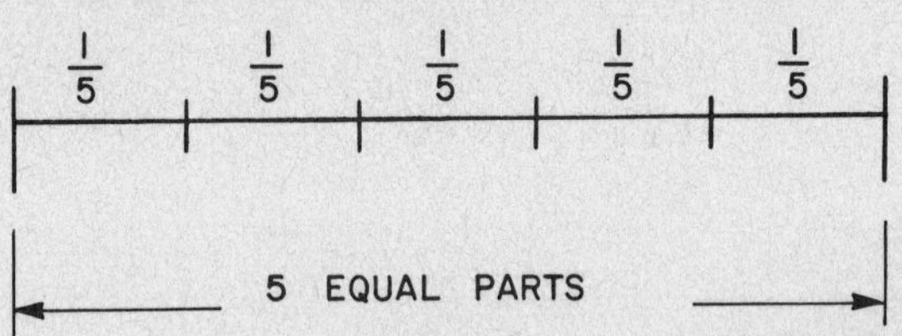

Figure 2-2

The denominator of a common fraction indicates how many equal parts the whole has been divided into. The numerator indicates how many of those parts the fraction represents. This concept is illustrated in Figure 2-2.

In Figure 2-2, a line has been divided into five equal parts. Each segment of the line equals 1/5 of the whole. Two segments represent 2/5 of the whole. Three segments represent 3/5 of the whole. Four segments represent 4/5 of the whole. And, of course, five segments equal the whole.

EXPRESSING FRACTIONS AS EQUIVALENTS

If you divide a pie into two equal parts and take one, you will have the same amount of pie as you will have if you divide it into four equal parts and take two (Figure 2-3). In other words, 1/2 is the same or equivalent of 2/4. The fraction 1/2 has many other equivalent ways in which it can be expressed (e.g., 4/8, 5/10, 6/12, 7/14, 8/16, 9/18, . . .). In order to be able to add, subtract, multiply, and divide common fractions, you must first be able to express them in equivalent form.

$$\frac{3}{4} = \frac{12}{16}$$

The example above shows how 3/4 is expressed in 16ths. This expression was accomplished as follows:

Step 1

Set up the problem by placing the known fraction on the left side of an equals sign (=). Place the denominator of the equivalent to the right

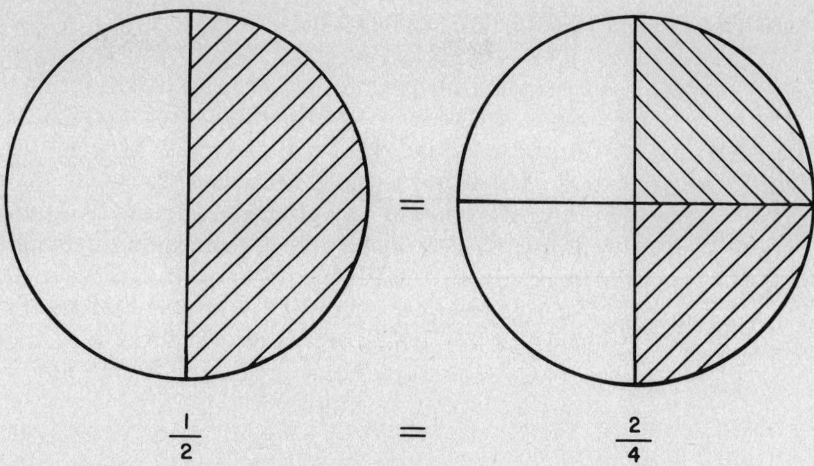

Figure 2-3

of the equals sign with a bar (–) over it. The numerator is left blank, of course, because it is the missing ingredient.

Step 2

Divide the denominator of the known fraction (4) into the denominator of the unknown equivalent (16). The result of this division is 4.

$$\frac{3}{4} = \frac{\quad}{16} \qquad \begin{array}{r} 4 \\ 4\overline{)16} \\ -\underline{16} \end{array}$$

Step 3

Multiply the quotient of the division step (4) times the numerator of the known fraction (3).

$$\frac{3}{4} = \frac{\quad}{16} \qquad 4 \times 3 = 12$$

Step 4

Write the product of the multiplication step (12) in the place left open for the numerator of the equivalent fraction.

$$\frac{3}{4} = \frac{12}{16}$$

PRACTICE PROBLEM SET 2-1

Express each fraction as an equivalent.

1. $\frac{3}{4} = \frac{?}{8}$
2. $\frac{1}{2} = \frac{?}{10}$
3. $\frac{5}{8} = \frac{?}{16}$
4. $\frac{3}{8} = \frac{?}{32}$
5. $\frac{9}{16} = \frac{?}{32}$
6. $\frac{1}{4} = \frac{?}{16}$
7. $\frac{7}{32} = \frac{?}{64}$
8. $\frac{3}{16} = \frac{?}{64}$
9. $\frac{5}{64} = \frac{?}{128}$
10. $\frac{3}{5} = \frac{?}{25}$

REDUCING FRACTIONS TO THE SIMPLEST TERMS

It is common practice to express fractions in the *simplest terms.* For example, 4/8 reduced to its simplest terms is 1/2; 12/16 reduced to its simplest terms is 3/4; and 8/32 reduced to its simplest terms is 1/4.

A fraction is in its simplest form when the numerator and denominator no longer contain any common factors. A *common factor* is a number that can be divided equally into both the numerator and the denominator. A factor common to both the numerator and the denominator of 12/16 is 4. Four will divide equally into 12 and 16. Performing these divisions will reduce 12/16 to 3/4, which is its simplest form.

Common fractions such as this one may be reduced to simplest form using a simple procedure:

$$\frac{15}{25} = \text{? in simplest form}$$

Step 1

Examine the numerator and the denominator to identify a common factor. In the example above, 5 is a common factor.

Step 2

Arrange the fraction so that you divide the numerator and denominator by the common factor (5).

$$\frac{15 \div 5}{25 \div 5} = \frac{3}{5}$$

Fifteen divided by 5 equals 3. Twenty-five divided by 5 equals 5. These two operations convert 15/25 to the simpler form 3/5.

Step 3

Check the numerator and denominator for another common factor. In this case there are no additional common factors. Therefore, 3/5 is the simplest form of 15/25.

It sometimes takes more than one set of divisions to reduce a fraction to its simplest form, especially if you are not experienced in reducing fractions. This situation is illustrated in the following steps:

$$\frac{9}{27} = \text{? in simplest form}$$

Step 1

Examine the numerator and denominator and identify a common factor. In the example above, 3 would be a common factor. Divide the numerator and denominator by 3. These divisions reduce the fraction to 3/9.

$$\frac{9 \div 3}{27 \div 3} = \frac{3}{9}$$

Step 2

Examine the numerator and denominator of the new fraction for common factors. In this case there is another common factor (3). Divide the numerator and denominator by 3. These divisions reduce the fraction to 1/3.

$$\frac{3 \div 3}{9 \div 3} = \frac{1}{3}$$

Since there are no additional common factors, 1/3 is 9/27 expressed in its simplest form.

This fraction (9/27) could have been reduced to its simplest form using just one set of divisions had we used 9 as a common factor, as in the following example:

$$\frac{9 \div 9}{27 \div 9} = \frac{1}{3}$$

Occasionally, a fraction may be presented in which the numerator is greater than the denominator:

$$\frac{5}{4} \quad \frac{31}{8} \quad \frac{75}{64}$$

Such fractions are *not* expressed in the simplest terms. The procedure for reducing these types of fractions to the simplest terms is simple:

Step 1

Subtract the denominator from the numerator.

$$5 \div 4 = 1 \text{ R1} \qquad 31 \div 8 = 3 \text{ R7} \qquad 75 \div 64 = 1 \text{ R11}$$

Step 2

The first number represents a whole number. The remainder is the numerator in the new fraction. The denominator remains the same.

1 1/4   3 7/8   1 11/64

The result is a mixed number. Mixed numbers are discussed in Chapter 3.

PRACTICE PROBLEM SET 2-2

Reduce each fraction to its simplest form.

1. 4/8	2. 3/9	3. 8/12	4. 2/6	5. 12/16
6. 12/32	7. 16/32	8. 32/64	9. 48/64	10. 64/30

ADDING COMMON FRACTIONS

It is important for technicians to understand how to add common fractions. This section provides a comprehensive review of the addition of common fractions.

Before attempting to add fractions, you must understand how to convert them to their least common denominator. Common fractions can be added only if they share the same denominator.

Adding Fractions with Like Denominators

Adding fractions that have like denominators is simple; just add numerators. In the following example, the answer (11/13) was obtained by adding the numerators 5 and 6.

$$5/13 + 6/13 = 11/13$$

Notice in the example above that the denominator does not change. Since 11/13 cannot be reduced to a simpler form, it is the final answer.

Adding fractions that do not have like denominators is not as simple as the example above, but it is not difficult. The first step is to determine the least common denominator of the fractions.

Determining Least Common Denominators

The following fractions cannot be added together in their present form:

$$3/4 + 1/16 = ?$$

However, if a common denominator for these fractions can be determined, adding them becomes simple. The least common denominator is the lowest number into which all denominators of fractions to be added can be divided equally.

The denominators of the fractions in the example above are 4 and 16. The least common denominator is 16, since 4 will divide evenly into 16 four times and 16 will divide evenly into 16 once. By converting 3/4 to the equivalent fraction 12/16, the fractions can be added simply by adding the numerators:

$$12/16 + 1/16 = 13/16$$

It is not always this simple to determine the least common denominator for fractions. However, there is a step-by-step procedure that can be used when the least common denominator is not readily apparent.

To determine difficult least common denominators, you have to understand three terms: factor, prime number, and prime factor. A *factor* is any number that is being multiplied. A *prime number* is any whole number other than 0 or 1 that is divisible only by itself or 1. For example, 3 is a prime number because it is divisible by only 3 or 1. Four is not a prime number because it is divisible by 2. A *prime factor* is any factor that is a prime number.

Factoring Denominators

Any whole number can be factored. This is the first step in determining least common denominators. To factor a denominator, write it as the product of prime factors (factors that are prime numbers). The following numbers have been factored using this method:

$$24 = 2 \times 2 \times 2 \times 3$$
$$18 = 2 \times 3 \times 3$$
$$16 = 2 \times 2 \times 2 \times 2$$
$$12 = 2 \times 2 \times 3$$
$$9 = 3 \times 3$$
$$8 = 2 \times 2 \times 2$$
$$4 = 2 \times 2$$
$$3 = \text{(prime number)}$$

Adding Fractions with Different Denominators

Once you have factored the denominators, you are ready to determine the least common denominator and add the fractions. This is done using the process of multiplication and addition.

Consider the following problem:

$$5/16 + 5/12 = ?$$

First, factor the denominators writing one factor string under the other:

$$16 = 2 \times 2 \times 2 \times 2$$
$$12 = 2 \times 2 \times 3$$

Now, count the *greatest* number of times that each prime factor is used and set up a factor string for multiplication.

The factor 2 is used four times in the first string and the factor 3 is used once in the second string. The resulting factor string is

$$2 \times 2 \times 2 \times 2 \times 3 = 48$$

By multiplying the string, you can determine that the lowest common denominator for 16 and 12 is 48.

The fractions are added by converting them to their equivalents in 48ths and adding numerators:

$$5/16 = 15/48 \qquad 5/12 = 20/48$$
$$15/48 + 20/48 = 35/48$$

PRACTICE PROBLEM SET 2-3

Complete each addition problem. Show all work. Express each answer in the simplest terms.

1. 1/3 + 2/3
2. 1/4 + 3/4
3. 3/8 + 5/8
4. 2/9 + 4/9
5. 1/2 + 3/4
6. 3/5 + 3/4
7. 1/6 + 5/9
8. 4/5 + 6/7
9. 3/24 + 7/18
10. 3/8 + 9/32

SUBTRACTING COMMON FRACTIONS

It is important for technicians to understand how to subtract common fractions. This section provides a comprehensive review of the subtraction of common fractions.

As with addition, to subtract fractions the fractions must have common denominators. Common fractions can be subtracted only if they share the same denominator. Once fractions have been expressed as equivalent with common denominators, the subtraction process is simple.

Step 1

Subtract the smaller numerator from the larger and write the answer over the common denominator.

$$7/12 - 1/12 = 6/12$$

Step 2

Reduce the answer to its simplest terms if necessary.

$$6/12 = 1/2$$

PRACTICE PROBLEM SET 2-4

Complete each subtraction problem. Show all intermediate steps. Express each answer in the simplest terms.

1. $3/4 - 1/4$
2. $7/8 - 5/8$
3. $15/64 - 7/64$
4. $45/132 - 13/132$
5. $7/8 - 3/4$
6. $11/12 - 2/3$
7. $5/9 - 1/6$
8. $6/7 - 4/5$
9. $3/8 - 9/32$
10. $7/18 - 3/24$

MULTIPLYING COMMON FRACTIONS

It is important for technicians to understand how to multiply common fractions. For example, to compute the area of a space dimensioned using fractions, technicians must be able to multiply common fractions.

You will recall from Chapter 1 that multiplication of whole numbers is a way to shortcut the addition process. The same is true of fractions. It is important to remember this when multiplying fractions. In this section we review how to multiply a whole number times a fraction and a fraction times a fraction.

Multiplying a Whole Number Times a Fraction

Multiplying a whole number times a fraction is the same as adding the fraction that number of times:

$$1/3 \times 3 = 1/3 + 1/3 + 1/3$$

The multiplication process shortcuts the addition process as follows:

Step 1

Multiply the whole number times the numerator and write the answer over the denominator.

$$7 \times 2/3 = 14/3$$

Step 2

If necessary, reduce the fraction to its simplest terms.

$$14/3 = 3\ 2/3$$

PRACTICE PROBLEM SET 2-5

Complete each multiplication problem. Express each answer in the simplest terms.

1. 5 × 2/3	2. 7 × 1/8	3. 3 × 3/4	4. 6 × 7/8
5. 12 × 1/12	6. 9 × 3/16	7. 14 × 1/4	8. 3 × 15/16
9. 11 × 5/8	10. 29 × 7/64		

Multiplying a Fraction Times a Fraction

Multiplying a fraction times a fraction is different from multiplying a whole number times a fraction, but it is a simple process.

$$1/3 \times 3/4 = 3/12 = 1/4$$

The problem in the example above is solved as follows:

Step 1

Multiply the numerators and write the product as the new numerator. Multiply the denominators and write the product as the new denominator.

$$2/3 \times 7/8 = 14/24$$

Step 2

If necessary, reduce the fraction to its simplest terms.

$$14/24 = 7/12$$

PRACTICE PROBLEM SET 2-6

Complete each multiplication problem. Express each answer in the simplest terms.

1. 1/4 × 1/4	2. 2/3 × 3/4	3. 7/8 × 1/4	4. 6/7 × 5/9
5. 4/5 × 9/10	6. 15/16 × 1/3	7. 5/12 × 1/2	8. 7/64 × 9/32
9. 19/132 × 15/16	10. 13/16 × 5/32		

DIVIDING COMMON FRACTIONS

It is important for technicians to understand how to divide common fractions.

You will recall from Chapter 1 that the division of whole numbers is a way to shortcut the subtraction process. The same is true of fractions. Consider this example of a division problem:

$$3/4 \div 3/8 = ?$$

This problem asks: "How many times will 3/8 fit into 3/4?" One way to answer this question is by subtracting 3/8 from 3/4 until you reach 0.

$$3/4 = 6/8$$

$$6/8 - 3/8 = 3/8 \quad \text{one subtraction}$$

$$3/8 - 3/8 = 0 \quad \text{two subtractions}$$

It took two subtractions to reach 0. Consequently, $3/4 \div 3/8 = 2$. This is the long way. It can be shortened by dividing. To do so you need to know how to (1) divide a fraction by a whole number, (2) divide a whole number by a fraction, and (3) divide a fraction by a fraction.

Dividing a Fraction by a Whole Number

Division of a fraction by a whole number is accomplished by inverting the divisor and multiplying.

$$1/4 \div 10 = 1/4 \times 1/10 = 1/40$$

Notice that when dividing a fraction by a whole number, a 1 is used as a space saver in the numerator after the inversion. A step-by-step explanation of this process follows.

Step 1

Rewrite the problem with the divisor inverted. Use a 1 for the numerator of the whole number.

$$1/4 \div 10 = 1/4 \times 1/10$$

Step 2

Multiply the numerators. Multiply the denominators. Reduce the answer to the simplest terms if necessary.

$$1/4 \times 1/10 = 1/40$$

PRACTICE PROBLEM SET 2-7

Complete each division problem. Express each answer in the simplest terms.

1. $2/3 \div 5$	2. $1/4 \div 3$	3. $9/16 \div 6$	4. $3/4 \div 4$	5. $11/15 \div 2$
6. $6/7 \div 7$	7. $1/2 \div 3$	8. $2/5 \div 2$	9. $7/9 \div 12$	10. $7/64 \div 9$

Dividing a Whole Number by a Fraction

Division of a whole number by a fraction is accomplished by writing a denominator of 1 under the whole number, inverting the divisor, and multiplying.

$$5 \div 1/4 = 5/1 \times 4/1 = 20/1 = 20$$

A step-by-step explanation of this process follows.

Step 1

Rewrite the problem with the divisor inverted. Use a 1 for the denominator of the whole number.

$$5 \div 1/4 = 5/1 \times 4/1 =$$

Step 2

Multiply the numerators. Multiply the denominators. Reduce the answer to the simplest terms, if necessary.

$$5/1 \times 4/1 = 20/1 = 20$$

PRACTICE PROBLEM SET 2-8

Complete each division problem. Express each answer in the simplest terms.

1. $5 \div 2/3$	2. $3 \div 1/4$	3. $6 \div 9/16$	4. $4 \div 3/4$	5. $2 \div 11/15$
6. $6 \div 6/7$	7. $3 \div 1/2$	8. $2 \div 2/5$	9. $12 \div 7/9$	10. $9 \div 7/64$

Dividing a Fraction by a Fraction

Dividing a fraction by a fraction is accomplished by inverting the divisor and multiplying.

$$3/4 \div 3/8 = 3/4 \times 8/3 = 24/12 = 2$$

A step-by-step explanation of this process follows.

Step 1

Rewrite the problem with the divisor inverted.

$$3/4 \div 3/8 = 3/4 \times 8/3$$

Step 2

Multiply the numerators. Multiply the denominators. Reduce the answer to the simplest terms.

$$3/4 \times 8/3 = 24/12 = 2$$

PRACTICE PROBLEM SET 2-9

Complete each division problem. Express each answer in the simplest terms.

1. $1/2 \div 1/4$	2. $1/3 \div 1/2$	3. $2/3 \div 2/5$	4. $3/4 \div 1/6$	5. $11/12 \div 7/8$
6. $5/5 \div 3/4$	7. $8/9 \div 2/3$	8. $7/8 \div 2/3$	9. $1/4 \div 5/7$	10. $3/64 \div 1/32$

REVIEW PROBLEMS

Express each fraction as an equivalent.

1. $\frac{7}{9} = \frac{?}{63}$ 2. $\frac{1}{5} = \frac{?}{125}$ 3. $\frac{7}{8} = \frac{?}{56}$ 4. $\frac{3}{8} = \frac{?}{64}$

Reduce each fraction to its simplest form.

5. $\frac{6}{8}$ 6. $\frac{56}{64}$

Complete each addition problem. Show all work. Express each answer in the simplest terms.

7. $\frac{3}{4} + \frac{3}{4}$ 8. $\frac{2}{3} + \frac{3}{5}$

Complete each subtraction problem. Show all intermediate steps. Express each answer in the simplest terms.

9. $\frac{7}{18} - \frac{1}{9}$ 10. $\frac{6}{11} - \frac{2}{33}$

Complete each multiplication problem. Express each answer in the simplest terms.

11. $8 \times \frac{2}{3}$ 12. $12 \times \frac{1}{2}$ 13. $\frac{2}{3} \times \frac{1}{2}$ 14. $\frac{5}{11} \times \frac{1}{3}$

Complete each division problem. Express each answer in the simplest terms.

15. $\frac{1}{3} \div 2$ 16. $\frac{7}{8} \div 6$ 17. $8 \div \frac{1}{2}$

18. $11 \div \frac{22}{3}$ 19. $\frac{1}{3} \div \frac{2}{5}$ 20. $\frac{5}{7} \div \frac{2}{3}$

3
MIXED NUMBERS

Having reviewed whole numbers and common fractions, you are now ready to move into the next phase of your arithmetic review. This phase deals with a review of mixed numbers. A mixed number is a whole number and a fraction. The following are mixed numbers:

5 1/3   17 9/64   9 15/16

Technicians must be proficient in the use of mixed numbers. To assist you in recalling your understanding of mixed numbers, this chapter covers the following major concepts: adding mixed numbers, subtracting mixed numbers, multiplying mixed numbers, and dividing mixed numbers.

ADDING MIXED NUMBERS

Mixed numbers are added by determining a common denominator for the fractions, rewriting the fractions as equivalents, adding the whole numbers, adding the fractions, and reducing the fraction to the simplest terms. A step-by-step explanation of how such a problem would be solved follows.

6 1/3 + 10 4/5 = ?

Step 1

Determine the least common denominator for the fractions and rewrite the problem with equivalent fractions.

6 5/15 + 10 12/15 = ?

Step 2

Add the whole numbers. Add the fractions. Reduce the answer to the simplest terms.

$$16\ 17/15 = 17\ 2/15$$

PRACTICE PROBLEM SET 3-1

Complete each addition problem. Express each answer in the simplest terms.

1. 2 1/3 + 2 2/3
2. 6 + 7 5/8
3. 5 3/4 + 4 1/8
4. 10 7/32 + 10 31/64
5. 67 1/2 + 15 3/32
6. 9 6/7 + 13 15/28
7. 4 8/25 + 9 7/50
8. 75 4/5 + 11 2/3
9. 4 13/45 + 3 1/15
10. 2 1/2 + 3 1/5

SUBTRACTING MIXED NUMBERS

Mixed numbers are subtracted by determining a common denominator for the fractions, rewriting the fractions as equivalents, subtracting the whole numbers, subtracting the fractions, and reducing the fraction to the simplest terms. A step-by-step explanation of how mixed numbers are subtracted follows.

$$10\ 4/5 - 6\ 1/3 = ?$$

Step 1

Determine the least common denominator for the fractions and rewrite the problem with equivalent fractions.

$$10\ 12/15 - 6\ 5/15 =$$

Step 2

Subtract the whole numbers. Subtract the fractions. Reduce the answer to the simplest terms.

$$10\ 12/15 - 6\ 5/15 = 4\ 7/15$$

When subtracting mixed numbers, the smaller mixed number may have a greater fraction than the larger mixed number, as in the following example:

$$10\ 1/3 - 5\ 3/4 = 10\ 4/12 - 5\ 9/12$$

When this happens, borrow a 1 from the whole number, convert it to its fractional equivalent, and add it to the smaller fraction.

$$10\ 4/12 \text{ after borrowing} = 9\ 16/12$$

The normal subtraction process may now be used.

$$9\ 16/12 - 5\ 9/12 = 4\ 7/12$$

PRACTICE PROBLEM SET 3-2

Complete each subtraction problem. Express each answer in the simplest terms.

1. 2 2/3 − 2 1/3	2. 7 5/8 − 2 1/4	3. 5 3/4 − 4 1/8
4. 10 31/64 − 10 7/32	5. 67 1/2 − 15 3/32	6. 13 15/28 − 9 6/7
7. 9 7/50 − 4 8/25	8. 75 4/5 − 11 2/3	9. 4 13/45 − 3 1/15
10. 3 1/5 − 2 1/2		

MULTIPLYING MIXED NUMBERS

Mixed numbers are multiplied by converting them to fractional equivalents, multiplying numerators, multiplying denominators, and expressing the result in the simplest terms. A step-by-step explanation of how mixed numbers are multiplied follows.

5 3/4 × 2 3/8 = ?

Step 1

Convert the mixed numbers to fractional equivalents by multiplying the denominator times the whole number and adding the numerator. The resulting number becomes the numerator for the mixed number's fractional equivalent. The denominator remains the same. Rewrite the problem.

5 3/4 = 23/4   2 3/8 = 19/8

23/4 × 19/8 = ?

Step 2

Multiply numerators. Multiply denominators. Express the result in the simplest terms.

23/4 × 19/8 = 437/32 = 13 21/32

PRACTICE PROBLEM SET 3-3

Complete each multiplication problem. Express each answer in the simplest terms.

1. 1 1/2 × 2 3/8	2. 2 1/3 × 4 1/8	3. 4 1/4 × 9 7/16
4. 11 4/5 × 2 3/8	5. 5 1/5 × 6 1/3	6. 8 7/8 × 4 1/2
7. 3 9/16 × 3 9/15	8. 1 1/8 × 4 1/8	9. 15 2/3 × 16 1/2
10. 7 3/4 × 1 1/2		

DIVIDING MIXED NUMBERS

Mixed numbers are divided by converting them to fractional equivalents, inverting the divisor, multiplying, and expressing the result in the simplest terms. A step-by-step explanation of how mixed numbers are divided follows.

6 7/8 ÷ 4 1/2 = ?

Step 1

Convert the mixed numbers to fractional equivalents and rewrite the problem.

$$55/8 \times 9/2 = ?$$

Step 2

Invert the divisor. Multiply the numerators. Multiply the denominators. Express the result in the simplest terms.

$$55/8 \times 2/9 = 110/72 = 1\ 38/72 = 1\ 19/36$$

PRACTICE PROBLEM SET 3-4

Complete each division problem. Express each answer in the simplest terms.

1. 2 3/8 ÷ 1 1/2
2. 4 1/8 ÷ 2 1/3
3. 9 7/16 ÷ 4 1/4
4. 11 4/5 ÷ 2 3/8
5. 6 1/3 ÷ 5 1/5
6. 8 7/8 ÷ 4 1/2
7. 8 9/16 ÷ 3 9/15
8. 4 1/8 ÷ 1 1/8
9. 16 1/2 ÷ 3 3/4
10. 7 3/4 ÷ 1 1/2

REVIEW PROBLEMS

Complete each addition problem. Express each answer in the simplest terms.

1. 6 1/2 + 6 1/2
2. 5 3/5 + 10 1/2
3. 7 3/7 + 3 1/2
4. 2 3/8 + 1 1/2
5. 3 1/3 + 3 3/4

Complete each subtraction problem. Express each answer in the simplest terms.

6. 11 1/2 − 9 1/4
7. 3 1/2 − 1 3/11
8. 9 7/8 − 5 1/6
9. 15 1/3 − 7 1/8
10. 5 4/9 − 4 1/2

Complete each multiplication problem. Express each answer in the simplest terms.

11. 15 1/8 × 4 1/2
12. 7 1/4 × 4 1/8
13. 11 5/9 × 2 1/3
14. 3 4/7 × 1 1/2
15. 2 6/7 × 7 2/7

Complete each division problem. Express each answer in the simplest terms.

16. 7 3/4 ÷ 4 1/2
17. 11 4/5 ÷ 1 1/2
18. 6 1/3 ÷ 5 1/5
19. 8 7/8 ÷ 1 1/8
20. 9 7/16 ÷ 3 9/15

4
DECIMAL FRACTIONS

Having reviewed whole numbers, common fractions, and mixed numbers, you are ready to move into the next phase of your arithmetic review. This phase deals with decimal fractions. Since one of the primary advantages of decimal fractions is their convenient use with digital calculators, you are encouraged to use a calculator in this and all subsequent chapters.

Decimal fractions are fractions with denominators that are *powers of 10.* Powers of 10 are numbers that result from multiplying 10 by itself a specified number of times. For example:

$$10 \times 10 = 100 \quad \text{(10 to the second power)}$$
$$10 \times 10 \times 10 = 1{,}000 \quad \text{(10 to the third power)}$$
$$10 \times 10 \times 10 \times 10 = 10{,}000 \quad \text{(10 to the fourth power)}$$

When using decimal fractions, a decimal point replaces the numerator-slash bar-denominator concept used with common fractions. For example:

Common Fraction		*Decimal Fraction*
1/10	=	0.10
1/100	=	0.01

Technicians must be proficient in the use of decimal fractions. Decimal fractions are used more than common fractions because they work conveniently with the popular digital calculators and computers.

To assist you in recalling your understanding of decimal fractions, this chapter covers the following major concepts: interpreting decimal fractions, converting common fractions to decimal fractions, converting decimal fractions to common fractions, adding decimal fractions, subtracting decimal fractions, multiplying decimal fractions, and dividing decimal fractions.

INTERPRETING DECIMAL FRACTIONS

While the denominator of a common fraction can be expressed as any number (e.g., 1/25, 1/3, 1/4, 1/67, 1/18, . . .), decimal fractions must be expressed in powers of 10 (e.g., 10, 100, 1,000, 10,000, 100,000, . . .). This means that with decimal fractions, one full unit is divided into 10, 100, 1,000, 10,000, 100,000, . . . parts. The fraction's part of the whole is expressed in the appropriate power of 10 (Figure 4-1). This line could have been divided into 100, 1,000, 10,000, 100,000, . . . parts. Each part would then be expressed in terms of this power of 10. For example:

1/100 =	0.01	(1 part of a line divided into 100 equal parts)
1/1,000 =	0.001	(1 part of a line divided into 1,000 equal parts)
1/10,000 =	0.0001	(1 part of a line divided into 10,000 equal parts)
1/100,000 =	0.00001	(1 part of a line divided into 100,000 equal parts)

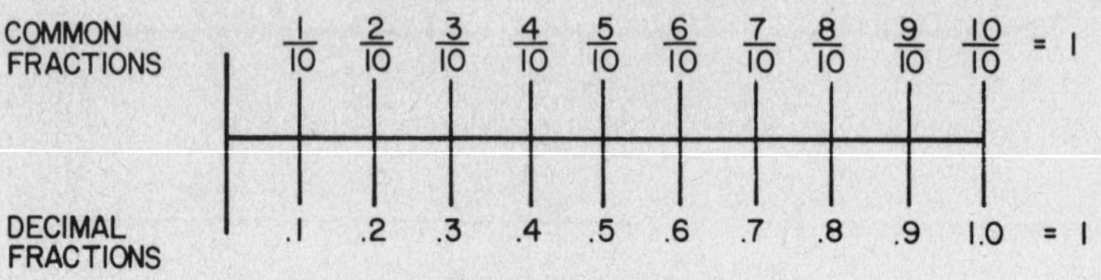

Figure 4-1

The number of digits to the right of the decimal in a decimal fraction determines the power of 10 being used in expressing the fraction. To read a decimal fraction, use a "place" chart similar to the one used for reading whole numbers in Chapter 1 (Figure 4-2).

The decimal fraction in this figure is read: "five hundred sixty-three thousand, seven hundred twenty-one millionths." The numbers to the right of the decimal point are read as a whole number as the numerator and the value of the farthest number to the right is stated as the denominator. Using this method yields the following equivalents:

0.563,721 = 563,721/1,000,000

The same procedure can be used for reading a whole number and a decimal fraction (mixed decimal):

56.12
35.004
46.1153

Using the procedure above for reading the decimal-fraction portion of the mixed decimal will produce the following interpretations:

56.12	fifty-six and twelve hundredths
35.004	thirty-five and four thousandths
46.1153	forty-six and one thousand one hundred fifty-three ten-thousandths

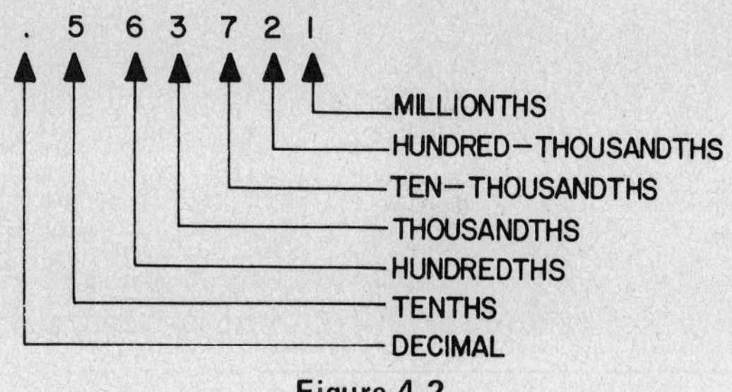

Figure 4-2

PRACTICE PROBLEM SET 4-1

Write out the word interpretation for each decimal fraction or mixed decimal.

1. 0.1	2. 0.5	3. 0.07	4. 0.113
5. 0.1456	6. 2.15	7. 6.892	8. 25.376
9. 325.4110	10. 1,412.969		

CONVERTING COMMON FRACTIONS TO DECIMAL FRACTIONS

The convenience, speed, and accuracy of digital calculators have increased the use of decimal fractions. However, common fractions are still used in various applications. For this reason, technicians must be able to convert common fractions to their equivalent decimal forms.

This is accomplished by dividing the numerator by the denominator, writing the result to the right of the decimal point, and rounding off to the specified number of places.

$$2/3 = 0.6666667 = 0.67$$

The conversion above was accomplished as follows:

Step 1

Rewrite the common fraction as a division problem with the numerator as the dividend and the denominator as the divisor.

$$3\overline{)2}$$

Step 2

Complete the division and round off the answer to two places.

$$3\overline{)\,2\,}\ \ 0.6666667 = 0.67$$

Step 2 required a new process, *rounding off.* It is sometimes necessary to round-off decimal fractions to shorten an answer. When this is the case, a simple process is used.

$$0.6666667 = 0.67$$

The example was rounded off from seven places to two places as follows:

Step 1

Count the specified number of places to the right of the decimal point and make a mental note of the number (6).

Step 2

Check the first number to the right of the 6 to determine if it is 5 or greater. If it is 5 or greater, round it up by 1 and drop the remaining numbers. Since the number in this case is 6, the number preceding it is changed to 7 and the remaining numbers are dropped. Had the number been less than 5, the remaining numbers would have been dropped and the 6 would have remained with no change of value.

Both of these possibilities are illustrated in the following examples, which have all been rounded off to three places:

$$0.712462 = 0.712$$
$$0.823344 = 0.823$$
$$0.897777 = 0.898$$
$$0.651413 = 0.651$$
$$0.100812 = 0.101$$

PRACTICE PROBLEM SET 4-2

Convert each common fraction to a decimal fraction. Round off each answer to two places.

1. 3/4	2. 2/3	3. 5/9	4. 3/64	5. 1/32
6. 5 3/8	7. 10 1/16	8. 112 5/7	9. 18 1/8	10. 153 4/11

CONVERTING DECIMAL FRACTIONS TO COMMON FRACTIONS

A common occurrence for technicians is to have to convert decimal fractions taken from a calculator to common fractions. This is accomplished by writing the decimal number as the numerator over the appropriate power of 10 as the denominator and reducing the result to the simplest terms.

$$0.05 = 5/100 = 1/20$$

The example above was converted according to the following steps:

Step 1

Read the decimal fraction as a whole number:

five one-hundredths

Step 2

Write the "five" as the numerator and the "one-hundredths" as the denominator. Reduce the common fraction to the simplest terms.

$$5/100 = 1/20$$

PRACTICE PROBLEM SET 4-3

Convert each decimal fraction to a common fraction. Express each answer in the simplest terms.

1. 0.73 2. 0.69 3. 0.55 4. 0.07 5. 0.04
6. 6.39 7. 11.08 8. 113.82 9. 19.15 10. 164.48

ADDING DECIMAL FRACTIONS

Adding decimal fractions is similar to adding whole numbers except that there is the decimal point. The key in adding decimal fractions is to align the decimal points when stacking them for addition.

$$\begin{array}{r} 97.35 \\ 21.627 \\ +\ \underline{34.5} \\ 153.477 \end{array}$$

The sample addition problem above is solved as follows:

Step 1

Stack the numbers to be added in the same manner as whole numbers, making sure that decimal points are aligned vertically.

$$\begin{array}{r} 97.35 \\ 21.627 \\ +\ \underline{34.5} \end{array}$$

Step 2

Add zeros to fill in empty spaces to the right of the decimal.

$$\begin{array}{r} 97.350 \\ 21.627 \\ +\ \underline{34.500} \end{array}$$

Step 3

Add each column of numbers using normal addition procedures. Bring the decimal point directly down from its position in the number stack.

$$\begin{array}{r} 97.350 \\ 21.627 \\ +\ \underline{34.500} \\ 153.477 \end{array}$$

PRACTICE PROBLEM SET 4-4

Complete each addition problem.

1. $\begin{array}{r} 62.470 \\ +\ \underline{3.568} \end{array}$

2. $\begin{array}{r} 12.92 \\ +\ \underline{13.41} \end{array}$

3. $\begin{array}{r} 123.456 \\ +\ \underline{55.555} \end{array}$

4. $\begin{array}{r} 414.612 \\ +\ \underline{333.753} \end{array}$

5. $\begin{array}{r} 1{,}415.92 \\ +\ \underline{538.05} \end{array}$

6. $\begin{array}{r} 8{,}794.100 \\ +\ \underline{9{,}675.742} \end{array}$

7. $\begin{array}{r} 54.92 \\ 67.69 \\ +\ \underline{81.63} \end{array}$

8. $\begin{array}{r} 71.342 \\ 27.369 \\ +\ \underline{19.182} \end{array}$

9. $\begin{array}{r} 359.386 \\ 621.404 \\ +\ \underline{892.403} \end{array}$

10. $\begin{array}{r} 1{,}984.23 \\ 851.39 \\ +\ \underline{4{,}999.81} \end{array}$

SUBTRACTING DECIMAL FRACTIONS

Subtracting decimal fractions is similar to subtracting whole numbers. As in adding them, the key in subtracting decimal fractions is aligning the decimal points when stacking the numbers for subtraction.

$$\begin{array}{r} 65.3213 \\ -\ \underline{24.662} \\ 40.6593 \end{array}$$

The sample subtraction problem above is solved as follows:

Step 1

Stack the numbers to be added in the same manner as whole numbers, making sure that decimal points are aligned vertically.

$$\begin{array}{r} 65.3213 \\ -\ \underline{24.662} \end{array}$$

Step 2

Add zeros to fill in empty spaces to the right of the decimal.

$$\begin{array}{r} 65.3213 \\ -\ \underline{24.6620} \end{array}$$

Step 3

Subtract each column of numbers using normal subtraction procedures. Bring the decimal point directly down from its position in the number stack.

$$\begin{array}{r} 65.3213 \\ -\ \underline{24.6620} \\ 40.6593 \end{array}$$

PRACTICE PROBLEM SET 4-5

Complete each subtraction problem.

1. 12.05 − 9.78
2. 28.16 − 14.432
3. 36.789 − 35.99
4. 887.42 − 96.98
5. 342.64 − 286.469
6. 550.01 − 461.005
7. 1,614.912 − 888.888
8. 1,792.1 − 1,513.28
9. 6,785.13 − 5,552.187
10. 10,865.01 − 8,621.192

MULTIPLYING DECIMAL FRACTIONS

Multiplying decimal fractions is similar to multiplying whole numbers. The major difference is that in multiplying decimal fractions you must count places to the right of the decimal point in order to place the decimal point properly in the final answer.

$$\begin{array}{r} 75.75 \\ \times\ \underline{4.38} \\ 331.785 \end{array}$$

The sample multiplication problem above is solved as follows:

Step 1

Stack the numbers to be multiplied, making sure that the digits line up beginning with the digits farthest to the right. Decimal points do *not* have to line up.

$$\begin{array}{r} 75.75 \\ \times \underline{\ \ 4.38} \end{array}$$

Step 2

Count the number of digits (places) to the right of the decimal in each number.

$$\begin{array}{rl} 75.75 & \text{(two places)} \\ \times \underline{\ \ 4.38} & \text{(two places)} \end{array}$$

Step 3

Multiply the numbers using normal multiplication procedures. Total the number of places to the right of the decimal from step 2 (four places). Place the decimal point to the left of the *fourth* digit from the right in the final answer.

$$\begin{array}{r} 75.75 \\ \times \underline{\ \ 4.38} \\ 60600 \\ 22725 \\ \underline{30300} \\ 331.7850 \end{array}$$

PRACTICE PROBLEM SET 4-6

Complete each multiplication problem. Round-off each answer to three places.

1. 15.06 × 36.32	**2.** 9.75 × 8.64	**3.** 16.51 × 18.89
4. 79.99 × 112.03	**5.** 342.65 × 66.72	**6.** 458.11 × 458.11
7. 0.341 × 0.769	**8.** 0.505 × 0.1143	**9.** 0.18 × 0.999
10. 0.001 × 0.501		

DIVIDING DECIMAL FRACTIONS

Dividing decimal fractions is similar to dividing whole numbers, but addition steps are required to contend with the decimal points.

$$0.562 \div 0.38 = 1.4789474 = 1.479$$

The sample division problem above is solved as follows:

Step 1

Rewrite the problem in the long-division format.

$$0.38\overline{)0.562}$$

Step 2

Move the decimal point in the divisor to the right of the right-hand-most number (two places). Move the decimal point in the dividend the same number of spaces (two places).

$$038.\overline{)056.2}$$

Step 3

Place the decimal point in the quotient directly above its position in the dividend and divide using normal division procedures.

```
       1.4
038./056.2
    - 38
      182
    - 152
       30
```

Step 4

Add zeros to the right of the last number in the dividend until the number divides with no remainder, proves to be a repeating decimal fraction (a decimal fraction that will never divide evenly, such as 0.666666), or you reach one digit more than the number of places to which you plan to round off.

```
       1.478 = 1.48
038./056.200
    - 38
      182
    - 152
       300
     - 266
        340
      - 304
         36
```

PRACTICE PROBLEM SET 4-7

Complete each division problem. Round-off each answer to two places.

1. $0.876 \div 0.36$
2. $10.59 \div 0.845$
3. $0.751 \div 0.621$
4. $14.616 \div 5.013$
5. $8.7 \div 0.1341$
6. $47.65 \div 13.10$
7. $155.12 \div 62.05$
8. $175.98 \div 39.72$
9. $13.65 \div 12.86$
10. $842.13 \div 94.88$

REVIEW PROBLEMS

Write out the word interpretation for each mixed decimal.

1. 101.9
2. 9.84
3. 1102

Convert each common fraction to a decimal fraction. Round-off each answer to two places.

4. 6/7 5. 11 2/3 6. 6 9/13

Convert each decimal fraction to a common fraction. Express each answer in the simplest terms.

7. 9.08 8. 19.811 9. 0.04

Complete each addition problem.

10. $\begin{array}{r} 54.93 \\ +\ \underline{359.387} \end{array}$ 11. $\begin{array}{r} 3.97 \\ 45.8 \\ +\ \underline{6.1} \end{array}$ 12. $\begin{array}{r} 6.124 \\ +\ \underline{3.542} \end{array}$

Complete each subtraction problem.

13. 37.16 − 9.78 14. 342.9 − 1.9 15. 1,674.9 − 666.66

Complete each multiplication problem. Round-off each answer to three places.

16. 0.341 × 36.1 17. 16.5 × 9.8 18. 55.5 × 0.8

Complete each division problem. Round-off each answer to two places.

19. 0.876 ÷ 0.37 20. 47.3 ÷ 13.1

5

POWERS, ROOTS, AND PERCENTAGES

Now that you have completed your review of whole numbers, common fractions, mixed numbers, and decimal fractions, you are ready to review powers, roots, and percentages. Technicians frequently deal with situations requiring the use of powers, roots, and percentages.

Powers are used when computing the area and volume. Roots are used in computing length, width, and height dimensions. Percentages are used in making a multitude of comparisons. Powers, roots, and percentages are reviewed in this chapter.

POWERS

A *power* is the product of two equal factors that have been multiplied. *Factors* are two or more numbers that are multiplied to produce another number. In determining the factors of a given number, you must ask yourself: "What numbers multiplied together will produce this number?" Some examples of numbers and factors of those numbers follow:

Number	*Factor*
10	2 and 5
8	2 and 4
6	2 and 3
4	2 and 2

Notice in the examples above that only the number 4 has *equal* factors. Numbers that have *equal* factors can be expressed as powers of those factors. For example, the number 4 can also be written as 2^2. This is read "2 to the second power" (or "2 squared"). The small raised 2 to the upper right of the 2 is called an *exponent.*

An exponent indicates how many multiplication operations are to take place. For example:

$$2^2 = 2 \times 2 = 4$$
$$2^3 = 2 \times 2 \times 2 = 8$$
$$2^4 = 2 \times 2 \times 2 \times 2 = 16$$

An exponent may be attached to both whole numbers as shown above, or decimal fractions as follows:

$$1.8^2 = 1.8 \times 1.8 = 3.24$$
$$2.9^3 = 2.9 \times 2.9 \times 2.9 = 24.389$$

From these examples it can be seen that 2^2 is another way to express 4, 2^4 is another way to express 16, and 2.9^3 is another way to express 24.389. When you determine what the corresponding number for a power and exponent is ($2^3 = 8$), the operation is called *raising.* You *raise* the number to the *power* indicated by the *exponent.* Raise 5 to the second power means perform the following operation:

$$5^2 = 5 \times 5 = 25$$

PRACTICE PROBLEM SET 5-1

Raise each number to the power indicated by the exponent.

1. 5^3	2. 1^4	3. 10^2	4. 15^3	5. 7^4
6. 1.65^2	7. 8.07^2	8. 9.25^3	9. 10.12^3	10. 17.2^4

ROOTS

When you have a number raised to a power and want to find the factor, you are looking for a *root.* A root is the opposite of a raised power. For example, 5 raised to the second power is 25. Correspondingly, the square root of 25 is 5. This is because 5 squared (5^2) equals 25. Remember that 5 to the second power (5^2) can also be read as "5 squared." Five to the third power can also be read "5 cubed." It is not common practice to carry this way of reading exponents beyond the third power. For example, 5^4 is usually only read as "5 to the fourth power."

In the preceding section you reviewed raising a number to a power. In this section you will review *taking the root* of a number, which is the opposite operation. The symbol for this operation is the *radical sign* ($\sqrt{}$). If you wish to take this square root of a number, it is common practice to use the radical sign by itself ($\sqrt{}$). If you wish to take the cube root or any root beyond (4, 5, 6, . . .), the root is indicated outside the radical sign:

$$\sqrt[3]{}$$
$$\sqrt[4]{}$$
$$\sqrt[5]{}$$

Taking the square root of a number is a task frequently performed by technicians. Taking the cube root is sometimes done. Taking roots beyond the third is uncommon.

In taking a root, you are asking yourself: "What number can be multiplied the number of times indicated by the root to equal the number under the radical sign?" For example, $\sqrt{9}$ asks, "What number can be multiplied by itself to equal 9?" The answer is 3. Therefore, 3 is the square root of 9. The operation $\sqrt[4]{16}$ asks, "What number can be multiplied 4 times to equal 16?" The answer is 2. Therefore, 2 is the 4th root of 16.

Taking roots used to be a long and difficult process alleviated only partially by special "root tables." However, the age of computers and calculators has solved this problem. Electronic calculators are equipped with a root function. It is recommended that you use such a calculator when taking roots.

PRACTICE PROBLEM SET 5-2

Take the indicated root in each problem. Use an electronic calculator with a root function.

1. $\sqrt{12}$
2. $\sqrt{17}$
3. $\sqrt{86.51}$
4. $\sqrt{194.06}$
5. $\sqrt{1{,}416.12}$
6. $\sqrt[3]{19}$
7. $\sqrt[3]{28.25}$
8. $\sqrt[3]{131.06}$
9. $\sqrt[3]{666.66}$
10. $\sqrt[3]{1{,}198.54}$

PERCENTAGES

Problems requiring the computation of percentages are very common. A list of all the many ways in which percentages are used would run on almost ad infinitum. In computing percentages, you are breaking a whole (any whole) into 100 parts and determining how many of those parts you have. One part equals 1 percent (written 1%). Two parts equal 2%, 10 parts equal 10%, and 78 parts equal 78%.

If you take a whole pie (100%) and divide it into four pieces, each piece equals 25% of the pie (Figure 5-1). This is determined by dividing the whole into the part. The whole in Figure 5-1 consists of four parts. One part divided by 4 (the whole) equals 25% or 0.25.

$$1/4 = 0.25 = 25\%$$
$$3/4 = 0.75 = 75\%$$
$$5/8 = 0.625 = 62.5\%$$

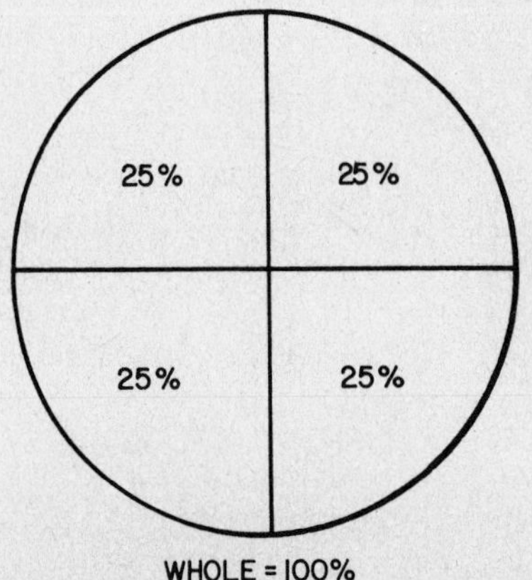

Figure 5-1

Notice in the examples above the relationship between a decimal fraction and an expressed percentage (0.25 = 25%, 0.625 = 62.5%). This relationship is based on the fact that 1% = 1/100 of a whole. The common fraction 1/100 is written 0.01 in decimal form. The decimal fraction 0.01 is expressed as a percentage by moving the decimal point two places to the right:

0.01 = 1. or 1%

The rule in converting decimal fractions to percentage expressions is:

Move the decimal point two places to the right.

0.001 = 0.1%
0.929 = 92.9%
0.0845 = 8.45%

The rule for converting common fractions to percentage expressions is:

Convert the common fraction to decimal form and move the decimal two places to the right.

3/16 = 0.1875 = 18.75%
4/5 = 0.8 = 80%
7/64 = 0.1094 = 10.94%

The rule for converting mixed numbers to percentage expressions is:

Let the whole number remain unchanged, convert the fraction to decimal form, and move the decimal two places to the right.

5 2/3 = 5.67 = 567%
19 7/8 = 19.875 = 1,987.5%
6 1/4 = 6.25 = 625%

PRACTICE PROBLEM SET 5-3

Express each fraction as a percent.

1. 0.0215	2. 0.5612	3. 0.0016	4. 9/11	5. 8/9
6. 15/64	7. 17/19	8. 3 3/8	9. 12 6/7	10. 14 11/13

Converting Percent Expressions to Decimal

Percent expressions may be converted to decimal numbers by dividing by 100. The division is actually accomplished by moving the decimal point two places to the left. This has the same effect as dividing by 100.

52% = 0.52
670% = 6.7

This operation can also be accomplished on percent expressions that involve whole numbers and fractions:

$$16\ 3/8\% = 16.375\% = 0.16375$$

The operation above is accomplished as follows:

Step 1

Rewrite the mixed whole number and fractional expressions as a decimal by dividing the fraction's denominator into its numerator and placing a decimal point between the whole number and your answer.

$$16\ 3/8\% = 16.375\%$$

Step 2

Move the decimal point two places to the left and drop the percent sign.

$$0.16375$$

PRACTICE PROBLEM SET 5-4

Convert each percent expression to a decimal.

1. 68%
2. 75%
3. 951%
4. 887%
5. 1,418%
6. 15 1/2%
7. 27 3/4%
8. 189 2/5%
9. 1 3/8%
10. 19 7/16%

Converting Percent Expressions to Common Fractions

Percent expressions can be converted to common fractions using a two-step procedure. In step 1, the percent expression is converted to decimal. In step 2, the decimal is converted to a common fraction.

$$16.2\% = 0.162 = 81/500$$

The operation above is accomplished as follows:

Step 1

Rewrite the percent expression as a decimal by moving the decimal point two places to the left.

$$16.2\% = 0.162$$

Step 2

Rewrite the decimal fraction as a common fraction and reduce to the simplest terms.

$$16.2\% = 162/1{,}000 = 81/500$$

PRACTICE PROBLEM SET 5-5

Write each percent expression as a common fraction.

1. 19.5%
2. 15.75%
3. 11.05%
4. 12.33%
5. 9.25%
6. 7.62%
7. 1.14%
8. 2.16%
9. 8.55%
10. 17.01%

SOLVING PERCENTAGE PROBLEMS

All percentage problems consist of three components: the base, the rate, and the percent. The *base* is the whole of which a part will be taken. The size of that part is determined by the rate. The *rate* is the percent of the whole that will be taken. The *percent* is the actual part that is taken from the whole. Examine the following problem:

$$12\% \text{ of } 60 = 7.2$$

The problem above is solved as follows:

Step 1

Convert the rate to decimal and rewrite the statement as a multiplication problem. The rate is to be multiplied times the base.

$$0.12 \times 60 = ?$$

Step 2

Multiply the rate times the base to determine the percent.

$$0.12 \times 60 = 7.2$$

This means that 7.2 is 12% of 60 or, said another way, 12% of 60 is 7.2.

PRACTICE PROBLEM SET 5-6

Solve each percentage problem.

1. 10% of 79
2. 11% of 92
3. 15% of 65
4. 19.3% of 157
5. 152% of 26
6. 111% of 1,412
7. 79.5% of 62
8. 0.45% of 22
9. 17.61% of 0.015
10. 18.75% of 1.621

REVIEW PROBLEMS

Raise each number to the power indicated.

1. 9.51^2
2. 19.03^3

Take the indicated root in each problem.

3. $\sqrt{57.061}$
4. $\sqrt[3]{1{,}962.45}$

Express each fraction as a percent.

5. 0.0562
6. 1.479
7. 5 3/16
8. 15 9/11

Convert each percent expression to a decimal.

9. 75%
10. 852%
11. 69.75%
12. 15.03%
13. 9 7/16%
14. 22 3/8%

Write each percent expression as a common fraction.

15. 21.5%
16. 13.33%
17. 1.16%
18. 17.02%

Solve each percentage problem.

19. 18.45% of 156
20. 19.01% of 13.7621

PART 2
Algebra, Geometry, and Trigonometry

6

ALGEBRA: SYMBOLISM, SIGNED NUMBERS, AND ALGEBRAIC OPERATIONS

INTRODUCTION TO PART 2

Part 2 is a presentation of the *essential* concepts of algebra, geometry, and trigonometry. It is important to stress the term "essential" here because one could have an entire book, in fact several books, on each of these math areas. However, Part 2 consists of eight chapters which cover those concepts of algebra, geometry, and trigonometry that the technician is most likely to need on the job.

There are no Pretests or Posttests in Part 2. It is assumed that learners have little or no background in algebra, geometry, and trigonometry. Consequently, all concepts covered are presented one at a time using a step-by-step approach that is simple to follow and easy to understand. Each concept presented is followed immediately by Practical Applications Problems that require the learner to use the concept in solving math problems. Each chapter concludes with a comprehensive set of Review Problems so that students can further reinforce their learning for the chapter before proceeding to the next chapter and new learning.

Chapters 6 and 7 cover the essentials of algebra with an emphasis on solving equations. It is critical that technicians be able to solve a wide variety of equations in order to perform their jobs. Learners completing Chapters 6 and 7 will develop a solid background in this area.

Chapters 8, 9, and 10 cover the essentials of geometry. Many rules of geometry are presented and should be referred to in solving the problems in these chapters as well as when attempting to complete the projects in Part 3.

Chapters 11, 12, and 13 cover the essentials of trigonometry.

Algebra is one of the most important branches of mathematics in that it allows technicians to solve problems that would be difficult or impossible using arithmetic. In algebra, letters are substituted for numbers to create general rules known as "formulas." Formulas can significantly expand the problem-solving capabilities of technicians who know how to apply them.

Algebra is a branch of mathematics just as arithmetic is a branch of mathematics. Algebra is actually an extension of arithmetic. Consequently, all of the procedures reviewed in Chapters 1 to 5 for solving arithmetic problems apply in solving algebra problems.

A study of algebra must be divided into general logical content groups that progress from the simple to the complex, each successive group building on the one that preceded it. In keeping with this approach, this chapter covers four major content groups: symbolism, signed numbers, operations, and powers and roots.

SYMBOLISM

Symbols are an integral part of algebra. Substituting letters for numbers is the foundation of algebra. In algebra regular numbers are called *arithmetic numbers.* Letters used to represent numbers are called *literal numbers.*

10, 20, 12, 16.5, 5 1/2	arithmetic numbers
a, b, C, D, e, F	literal numbers

Algebraic Expressions

Algebraic expressions are word statements converted into mathematical form. An algebraic expression has three components: arithmetic numbers, literal numbers, and signs of operation.

Word Statement. An employer plans to deduct 2 hours from an employee's weekly time card for repeated lateness. How many hours will the employee be paid for?

Algebraic Expression. $x - 2 = ?$ (x is the literal number, − is the sign of operation, and 2 is the arithmetic number).

In the algebraic expression, the literal number (x) represents the total number of hours the employee will work for the week. At this time the employer does not know how many that will be, so he or she uses the literal number (x). When the employee's timecard is finally totaled for the week, the total will be substituted for the literal number (x) and 2 hours will be subtracted.

Multiplication Symbolism and Expressions in Algebra

In arithmetic the times sign (X) is used to indicate the multiplication operation. However, in algebra the times sign (X) would be mistaken for a literal number (x), so multiplication operations must be stated differently.

In algebra, an arithmetic number multiplied times a literal number requires no sign of operation:

$$10 \text{ times } a = 10a$$
$$6 \text{ times } X \text{ times } Y = 6XY$$

An arithmetic number times an arithmetic number does not require a sign of operation to avoid confusion. In this case parentheses ( ) are substituted for the regular times sign (×):

$$10 \times 10 = 10(10) \quad \text{or} \quad (10)10 \quad \text{or} \quad (10)(10)$$

Order of Operations

An algebraic expression can be resolved to equal a certain value by substituting arithmetic numbers for literal numbers and performing the operations indicated. The order in which operations are performed is important. There are several rules that will assist you in determining the value of an algebraic expression. These rules are known as the *order of operations.*

1. Perform operations enclosed in parentheses or brackets or involving a fraction bar or a radical sign.
2. Perform operations involving powers and roots.
3. Perform operations involving multiplication and division.
4. Perform operations involving addition and subtraction.

PRACTICE PROBLEM SET 6-1

Convert each word statement to an algebraic expression.

1. Increase the given length by 5 inches.
2. The product of 10, X, and Y.
3. Subtract B from 35.
4. The product of 13 and X increased by 15.
5. The square of Y divided by 3 times 5.

Determine the value of each algebraic expression by substituting the arithmetic numbers given for the literal numbers. Be sure to follow the order of operations.

6. $X + 5 = ?$ — $X = 6$
7. $6A - 25 = ?$ — $A = 5$
8. $\dfrac{X^2}{Z+7} + (B - 3) = ?$ — $X = 10,\ Z = 3,\ B = 5$
9. $\dfrac{15X}{Y} + [3X - (26 + 2) - Y] = ?$ — $X = 12,\ Y = 5$
10. $\dfrac{A^2 + B^3 + 9}{4A - 2B} + B^2(AB - 5B) = ?$ — $A = 15,\ B = 23$

SIGNED NUMBERS

Numbers such as 5, 3, 2.50, 1 7/8, and 12.03 are *positive numbers.* This means that on a number line, they fall to the right of zero. On a number line, the positive direction is right and the negative direction is left. Positive numbers may be signed using the positive sign (+) or they may be written without a sign. A number written without a sign is considered a positive number.

Numbers that fall to the left of zero on a number line are called *negative numbers* and *must* be preceded by a negative sign (−). The following numbers are negative numbers: −10, −3.75, −67.52, and −2 1/3.

Figure 6-1 is an example of a number line. Notice that the origin point is zero and that lines extend right and left from zero. Those that fall along

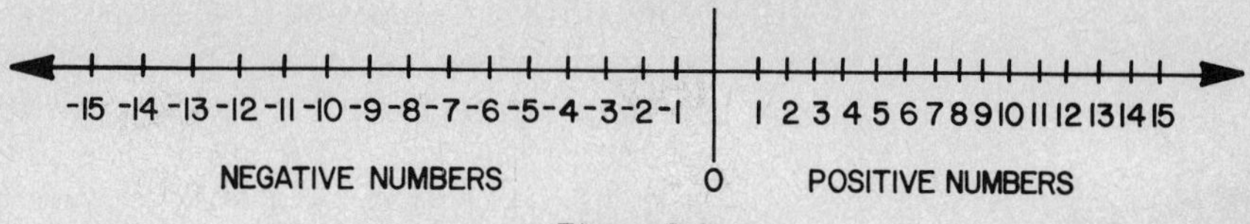

Figure 6-1

the extension of the line to the right are positive numbers. As is customary, they all appear without their corresponding positive (+) signs. Those that fall along the extension of the line to the left are negative. Negative numbers must be preceded by a negative (−) sign, as these are. Both extensions, to the right and to the left, run on infinitely.

In Chapter 1 you learned how to perform basic arithmetic operations on whole numbers. You can also perform these operations (addition, subtraction, multiplication, division, powers) on signed numbers. However, before learning to do so, you will need to learn a new concept—*absolute value.*

Absolute Value

The *absolute value* of a number is the number of units away from zero that it represents on a number line. An easier way to say this is that the absolute value of a number is its value without a sign. For example, the absolute value of −10 is 10. The absolute value of +10 is 10. This is because both numbers are 10 units away from zero on a number line (see Figure 6-1).

Numbers on the positive side of the number line have greater value than those on the left side. This is because those on the left side are *less than zero.* For example, a +1 is greater than a −15.

In addition, any number on the number line is considered greater than a number to its left. For example, +5 is greater than +4 and −5 is greater than −6.

These rules *do not* apply when dealing with the absolute value of numbers. "Absolute value" means to disregard relative locations on the number line and count only the number of units away from zero. For example, the absolute value of −5 is greater than the absolute value of +4. This is because 5 without a sign is greater than 4 with a sign.

Adding Signed Numbers

In adding signed numbers, you will face the following situations:

1. Adding positive numbers
2. Adding negative numbers
3. Adding positive and negative numbers

Positive numbers are added using the procedures learned in Chapter 1. Negative numbers follow the same procedure except that the *absolute values* are used and the sum is always negative, as in the following examples:

$$\begin{array}{r} -10 \\ \underline{-15} \\ -25 \end{array}, \quad \begin{array}{r} -132 \\ \underline{-104} \\ -236 \end{array}$$

$$-7 + (-4) + (-9) + (-3) = -23$$

In adding positive numbers, the sum may be positive or negative, depending on the absolute values of the numbers being added. Positive and negative numbers are added by *subtracting* the smaller absolute value from the larger and using the sign of the larger. For example:

$$5 + (-6) = -1$$
$$10 + (-4) = 6$$

In the first example, the absolute value of the negative number is greater (6). Therefore, the answer will be negative. The absolute value of 5 (5) subtracted from the absolute value of -6 (6) is 1. Consequently, the final answer is -1.

In the second example, the number with a greater absolute value is positive. Therefore, the answer will be positive. The absolute value of -4 (4) subtracted from the absolute value of 10 (10) is 6. Consequently, the final answer is 6.

The same procedures, with two steps added, can be used in adding several positive and negative numbers. For example:

$$-5 + 10 + (-6) + (-4) + 9 = 4$$

In this example, all positive numbers are added:

$$10 + 9 = 19$$

Then all negative numbers are added:

$$-5 + (-6) + (-4) = -15$$

Finally, the positive and negative numbers are added using the procedures described above:

$$19 + (-15) = 4$$

Subtracting Signed Numbers

To understand how to subtract signed numbers, you must know two rules:

1. Subtracting a negative number is the same thing as adding a positive number.
2. Subtracting a positive number is the same thing as adding a negative number.

The procedure for subtracting negative numbers involves two steps:

1. Change the sign of the subtracted to its opposite (positive to negative or negative to positive).
2. Add the numbers using the procedures described in the previous sections.

This two-step procedure is illustrated in the following examples:

$$10 - (-5) = 10 + 5 = 15$$
$$-15 - (+6) = -15 + (-6) = -21$$

In the first example, a negative 5 is to be subtracted from a positive 10. The negative 5 is changed to a positive 5 and the numbers are added using the normal procedure. In the second example, a positive 6 is to be subtracted from a negative 15. The positive 6 is changed to a negative 6 and the numbers are added using the normal procedure.

Multiplying Signed Numbers

Multiplying signed numbers is simple. It is accomplished by multiplying the absolute values of the numbers using normal multiplication procedures and adding a step. The additional step involves determining if the final answer is positive or negative.

The following rules of thumb will help in making this determination:

1. A negative times a negative equals a positive.
2. A positive times a positive equals a positive.
3. A negative times a positive equals a negative.
4. In a string, if all numbers are positive, the answer is positive.
5. If there is an even number of negative numbers, the final answer is positive.
6. If there is an odd number of negative numbers, the final answer is negative.

These rules are illustrated in the following examples:

$$-5 \times (-5) = 25$$
$$10 \times 12 = 120$$
$$-6 \times 3 = -18$$
$$5 \times 6 \times 2 \times 3 = 180$$
$$-4 \times 2 \times 3 \times 6 \times (-1) = 144$$
$$-3 \times 5 \times (-2) \times (-6) = -180$$

In the first example a negative times a negative produced a positive. In the second example, a positive times a positive produced a positive. In the third example, a negative times a positive produced a negative. In the fourth example, a string of positives produced a positive. In the fifth example, an even number of negatives (2) produced a positive. In the last example, an odd number of negatives (3) produced a negative.

Dividing Signed Numbers

Dividing signed numbers is a two-step procedure. In the first step, the absolute values are divided using normal division procedures. Then the sign of the final answer is determined. To make this determination, you need to know two rules:

1. If the divisor and dividend have the same sign, the quotient is positive.
2. If the divisor and dividend have different signs, the quotient is negative.

These rules were applied in solving the following problems:

$$10 \div 5 = 2$$
$$-12 \div (-6) = 2$$
$$-15 \div 3 = -5$$
$$20 \div (-5) = -4$$

In the first two examples, like signs produced a positive quotient. In the last two examples, unlike signs produced a negative quotient.

Raising Signed Numbers to a Power

Raising numbers to powers was covered in Chapter 5. Signed numbers with positive exponents can be raised to powers using those same procedures and applying the following rules:

1. A positive number raised to any power is positive.
2. A negative number raised to an odd power is negative.
3. A negative number raised to an even power is positive.

Raising signed numbers with positive exponents to powers is illustrated in the following examples:

$$5^2 = 25$$
$$-5^3 = -125$$
$$-5^2 = 25$$

In the first example a positive number raised to a power produced a positive answer. In the second example, a negative number raised to an odd power produced a negative answer. In the third example, a negative number raised to an even power produced a positive number.

PRACTICE PROBLEM SET 6-2

Write the absolute value of each number.

1. 5
2. −7
3. 3.4
4. 1 7/8
5. −66.8

Complete each addition problem.

6. 17 + (+7)
7. −8 + (−8)
8. 25 + (−15)
9. −30 + (+2)
10. 2 + 4 + 6 + (−1) + (−16)

Complete each subtraction problem.

11. 8 − (+3)
12. 5 − (−2)
13. −15 − (+10)
14. −34 − (−7)
15. 10.03 − (−3.43)

Complete each multiplication problem.

16. (−10)(5)
17. (−22)(−6)
18. 16(−2)
19. (−15.45)(0.05)
20. (−0.12)(−17.2)(6.42)

Complete each division problem.

21. −52 ÷ 5
22. −16 ÷ (−8)
23. 25 ÷ (−7)
24. −14.73 ÷ 3.25
25. −156.17 ÷ 14.73

Raise each signed number to the indicated power.

26. -5^2 27. -3^3 28. -2^4 29. -17.62^2 30. -15.91^3

ALGEBRAIC OPERATIONS

The basic operations of arithmetic are addition, subtraction, multiplication, division, and powers. These are also the basic operations of algebra. You must be able to perform basic algebraic operations in order to solve equations:

$$\frac{10 - 4X + 12XY - 5}{4} + \frac{5X^2(Y^4)}{16} = 156$$

The example above is an equation involving the basic algebraic operations of addition, subtraction, multiplication, division, and roots. To solve such an equation and perform the basic algebraic operations, you must first learn several concepts that are intrinsic to algebra. These factors include: algebraic expression, term, factor, numerical coefficient, literal factors, like terms, and unlike terms.

You learned earlier that an *algebraic expression* is a word statement expressed in mathematical terms using arithmetic numbers, literal numbers, and signs of operation. The left side of the equation above is an algebraic expression. The following examples are also algebraic expressions.

$$15x + 17xy - 5x^3y^2$$
$$9y + 8ab + 4cd$$

A *term* is any part of an algebraic expression separated by a plus or minus sign. In the first example above, $15x$, $17xy$, and $5x^3y^2$ are terms. In the second example, $9y$, $8ab$, and $4cd$ are terms.

A *factor* is part of a term. In the term $15x$, there are two factors: 15 and x. Factors of terms are multiplied together. Examine the following list of terms and their corresponding factors.

Term	*Factors*
$15xy$	15, x, and y
$9ab^2$	9, a, and b^2
$-5x^2y^2$	-5, x^2, and y^2

A term can have both number and letter factors. For example, the term $15x$ has a number factor (15) and a letter factor (x). The number factor is known as the *literal factor.*

In an algebraic expression, there may be like terms and unlike terms. *Like terms* are terms with common literal factors. *Unlike terms* do not have common literal factors.

$$7x + 12x - 17xy + 0.75x^2y^3$$

The expression above contains two terms with common literal factors ($7x$ and $12x$). These are like terms. It also contains two unlike terms ($17xy$ and $0.75x^2y^3$).

With this background you are ready to learn how to perform the basic algebraic operations of addition, subtraction, multiplication, division, powers, and roots.

Algebraic Addition

In algebraic expressions, only like terms can be added. The rules for adding signed numbers apply. This is accomplished by adding numerical coefficients and leaving the literal factors unchanged. For example:

$$15xy + 7xy = 22xy$$
$$-16X^2 + 8X^2 = -8X^2$$
$$3ab^2 + (-12ab^2) = -9ab^2$$

Unlike terms are *not* added. The addition is indicated by the sign, but numerical coefficients are *not* added.

Algebraic expressions containing several terms can be added together by first stacking and then adding like terms, as in the following example:

$$\begin{array}{rrrrr} & 15x + & 12xy + & 15.64x^2y^2 + & 4ab \\ + & 5x + & 9xy + & 3.51x^2y^2 + & 3ab \\ \hline & 20x + & 21xy + & 19.15x^2y^2 + & 7ab \end{array}$$

In the example above, like terms were stacked. Then the numerical coefficients of the like terms were added. Addition is indicated by the signs separating the unlike terms. At this point no further addition can be accomplished.

Algebraic Subtraction

In algebraic expressions, only like terms can be subtracted. This is accomplished by subtracting numerical coefficients and leaving the literal factors unchanged. For example:

$$15xy - 7xy = 8xy$$
$$-12ab - 0.56ab = -12.56ab$$
$$17.62x^2y^3 - 5.51x^2y^3 = 12.11x^2y^3$$

Unlike terms are *not* subtracted. The subtraction is indicated by the sign, but numerical coefficients are *not* subtracted.

Algebraic expressions containing several terms can be subtracted by first stacking and then subtracting like terms, as in the following example:

$$\begin{array}{rl} 10x - 15xy - 20xyz & = 10x - 15xy - 20xyz \\ -\ \underline{(9x + 6xy + 8xyz)} & = \underline{9x - 6xy - 8xyz} \\ & \quad x - 21xy - 28xyz \end{array}$$

In the example above, like terms were stacked. Then the expressions were rewritten to prepare for the subtraction process. Numerical coefficients of like terms were subtracted. Subtraction is indicated between unlike terms. At this point, no further subtraction can be accomplished.

Algebraic Multiplication

Algebraic multiplication is a simple two-step process and the rules for multiplying signed numbers apply. First, multiply each term in one expression times each term in the other expression. Then combine like terms.

$$(9x + 12y)(15xy - 3x^2y) = 135x^2y - 27x^3y + 180xy^2 - 36x^2y^2$$

The algebraic multiplication above was accomplished as follows:

1. $(9x)(15xy) = 135x^2y$
2. $(9x)(-3x^2y) = 27x^3y$
3. $(12y)(15xy) = 180xy^2$
4. $(12y)(-3x^2y) = -36x^2y^2$
5. $135x^2y - 27x^3y + 180xy^2 - 36x^2y^2$

At this point, no further multiplication can be accomplished.

When exponents are involved in a multiplication, they are added. For example:

$$(5X^4)(10X^2) = 50X^6$$
$$(15.56x^3y^2)(4x^2y^2) = 62.24x^5y^4$$

Algebraic Division

Algebraic division is a two-step process and the rules for dividing signed numbers apply. First, divide each term in the dividend by the divisor. Then combine like terms as in the following example:

$$\frac{10a^2b^2 + 12ab^3}{2ab}$$
$$5ab + 6b^2$$

The example problem above was solved as follows:

1. $\dfrac{10a^2b^2}{2ab} = 5ab$
2. $\dfrac{12ab^3}{2ab} = 6b^2$
3. $5ab + 6b^2$

In step 1, the numerical coefficient in the divisor (2) goes into the numerical coefficient in the dividend (10) five times. The literal factor a^2 divided by a equals a^1 or $a(a^{2-1})$. The literal factor b^2 divided by b equals b^1 or $b(b^{2-1})$.

In step 2, the numerical coefficient in the divisor (2) goes into the numerical coefficient in the dividend (12) six times. The literal factor a divided by a is 1. Consequently, it can be dropped since 1 multiplied times any factor causes no change. The literal factor b^3 divided by b is b^2 (b^{3-1}).

Algebraic Powers

Raising an algebraic expression to a power is a two-step process. First, rewrite the expression as a multiplication problem and perform the multiplication operations. Then combine like terms as in the following example.

$$(4a^2 - 3ab)^2 = 16a^4 - 24a^3b + 9a^2b^2$$

The example problem above was solved as follows:

1. $(4a^2 - 3ab)(4a^2 - 3ab) =$
2. $(4a^2)(4a^2) = 16a^4$
3. $(4a^2)(-3ab) = -12a^3b$
4. $(-3ab)(4a^2) = -12a^3b$
5. $(-3ab)(-3ab) = 9a^2b^2$
6. $16a^4 - 12a^3b + 9a^2b^2$

In step 1, the expression is rewritten as a multiplication problem. In steps 2 to 4, the multiplication operations are performed according to the rules for multiplying signed numbers. In step 5, there are no like terms to combine.

PRACTICE PROBLEM SET 6-3

Complete each addition problem.

1. $12ab + 5ab$
2. $5x + x$
3. $10x^2 + 15x^2$
4. $4.3x^2y^3 + 19x^2y^3$
5. $(5x + 9xy + 13x^2y) + (9xy + 5x^2y)$

Complete each subtraction problem.

6. $5ab - 2ab$
7. $10a^2b - 3a^2b$
8. $-6x - (+10x)$
9. $(16.45x^3y^4) - (+2.78x^3y^4)$
10. $(10xy + 5x^2y) - (15xy - 2x^2y)$

Complete each multiplication problem.

11. $(5x)(a)$
12. $10x^2(xy)$
13. $(-2ab)(6x)$
14. $(13.78a^5b^3)(-c^2)$
15. $(5x^2 + 9y)(4.5x - y)$

Complete each division problem.

16. $10a \div 2a$
17. $15x^2 \div 5x$
18. $6a^2b^2 \div 3a$
19. $(12c^3d^4) + 9ab \div 2c$
20. $7x^3 + 9x^2 \div 1.5x$

Raise each term to the power indicated.

21. $(2a + b)^2$
22. $(5x - y)^2$
23. $(10a)^3$
24. $(3xy)^2$
25. $(2a^2b)^2$

REVIEW PROBLEMS

Convert each word statement to an algebraic expression.

1. The product of 5 and X^2.
2. The cube of A divided by 6 times 2.

Determine the value of each algebraic expression by substituting arithmetic numbers for their corresponding literal numbers.

3. $\dfrac{X^2 + A^3}{12 + Y}$ $(X = 5, A = 6, Y = 10)$

4. $\dfrac{A^2 + B^2 + C^2}{15} + \dfrac{C^2 \times 6}{3}$ $(A = 3, B = 4, C = 5)$

Write the absolute values of each number.

5. 5 ____ ; -3 ____ ; -1.76 ____

6. 1 3/16 ____ ; -0.007 ____ ; 3.40 ____

Solve each signed number problem.

7. $19 + (-6)$
8. $35 + (+10)$
9. $120 - (-5)$
10. $-16 - (+9)$
11. $10(-7)$
12. $(-15.05)(0.07)$
13. $-47.49 \div 3.86$
14. $-39.01 \div 3.86$

Raise each number to the power indicated.

15. 52.41^2
16. 39.06^3

Perform each algebraic operation.

17. $6.9x^3y^2 + 29x^3y^2$
18. $(10ab + 6xy) + (14ab - 3x^2y^2)$
19. $(4y^2 + 3x)(3x)$
20. $16x^2 \div 2x$

7

ALGEBRA: SOLVING EQUATIONS

An equation is a statement of equality. An equation has three components:

1. The left side
2. The equal sign (=)
3. The right side

The fundamental rule for solving equations is *equality*. The left side of the equation equals the right side. No matter how one side of an equation is written, it must equal the other side. For example:

$$2x = 10$$
$$5x^2 + 4 = 152$$
$$(a)(c) = x - y$$

Many standard formulas used are equations. For example:

$$a^2 + b^2 = c^2$$
$$k = \frac{1}{2}ab$$
$$m = \frac{1}{2}(a + b)$$

Solving an equation involves determining the value of one or more unknowns. Unknowns are represented by letters:

$$2X + 5 = 15$$

In this equation the problem is to determine the value of X. The letter X represents the unknown. A rule of thumb for solving equations is:

Use the known quantities to solve for the unknown.

An equation is actually a word statement or problem stated in algebraic terms. Converting a word problem into an equation is the first step in solving the problem. What follows is an example of how this is done.

Word Problem

The latest version of a given piece of equipment costs 5% less than last year's model, which cost $38,000. What does the new version cost?

Step 1

Convert the problem to the shortest question possible.

What number is 95% of $38,000?

Step 2

Identify the unknown and assign a letter(s).

The only unknown in this problem is the *number* that is 95% of $38,000. Assign this number the letter X.

Step 3

Convert the question into an equation, substituting the letter for the unknown.

$$0.95X = 38{,}000.$$

At this point the equation is ready to be solved. There are several principles for solving equations that must be learned first:

1. The addition principle of equality
2. The subtraction principle of equality
3. The multiplication principle of equality
4. The division principle of equality
5. The root principle of equality
6. The power principle of equality

THE ADDITION PRINCIPLE OF EQUALITY

The addition principle is used when an equation is written such that a number is to be subtracted from the unknown. The *addition principle* states:

The same number can be added to both sides of an equation without altering the equality.

This principle gives you a tool that can be used in solving equations. It allows you to isolate the unknown on one side of the equation, which is the key to solving an equation. The addition principle is applied in solving for the unknown in the following equation:

$$X - 10 = 20$$
$$X - 10 + 10 = 20 + 10$$
$$X = 30$$

In the example above, 10 is subtracted from the unknown (X). The addition principle applies. Ten is added to both sides of the equation. On the left side of the equation, the $-$ 10 and $+$ 10 cancel each other out, isolating the unknown. On the right side, $20 + 10 = 30$. Therefore, $X = 30$.

PRACTICE PROBLEM SET 7-1

Solve each equation.

1. $X - 5 = 15$
2. $M - 3 = 6$
3. $A - 15 = 22.6$
4. $b - 6.73 = 12.46$
5. $Y - 0.003 = 3.75$
6. $19 = C - 7$
7. $156 = X - 5.51$
8. $47 = a - 4.16$
9. $13 = M - 2$
10. $33 = p - 17$

THE SUBTRACTION PRINCIPLE OF EQUALITY

The subtraction principle is used when a number is added to the unknown. The *subtraction principle* states:

The same number can be subtracted from both sides of an equation without altering the equality.

The subtraction principle is applied in solving for the unknown in the following equation:

$$X + 10 = 20$$
$$X + 10 - 10 = 20 - 10$$
$$X = 10$$

In the example above, 10 is added to the unknown. The subtraction principle applies. Ten is subtracted from both sides of the equation. On the left side of the equation, the $+$ 10 and $-$ 10 cancel each other out, isolating the unknown. On the right side, $20 - 10 = 10$. Therefore, $X = 10$.

PRACTICE PROBLEM SET 7-2

Solve each equation.

1. $X + 6 = 12$
2. $a + 3 = 15$
3. $b + 6.76 = 23.46$
4. $m + 1.19 = 42.76$
5. $p + 0.012 = 4.33$
6. $129 = c + 63.15$
7. $75 = q + 17$
8. $35 = v + 11.84$
9. $21 = y + 5.03$
10. $101.94 = T + 9$

THE MULTIPLICATION PRINCIPLE OF EQUALITY

The multiplication principle is used when a number is divided *into* the unknown. The *multiplication principle* states:

The same number can be multiplied times both sides of an equation without altering the equality.

The multiplication principle is applied in solving for the unknown in the following equation:

$$\frac{X}{10} = 20$$
$$\left(\frac{X}{10}\right)(10) = 20 \times 10$$
$$X = 200$$

In the example above, the unknown is divided by 10. The multiplication principle applies. Ten is multiplied times both sides of the equation. On the left side of the equation, the 10's cancel each other out, isolating the unknown. On the right side, $20 \times 10 = 200$. Therefore, $X = 200$.

PRACTICE PROBLEM SET 7-3

Solve each equation.

1. $X/5 = 50$
2. $M/4 = 12$
3. $P/3.25 = 6.72$
4. $Q/0.04 = 11.05$
5. $Y/4.69 = 22.46$
6. $16.03 = R/0.071$
7. $56.01 = A/0.92$
8. $13.55 = B/1.66$
9. $7.15 = C/14.11$
10. $9.63 = E/3.32$

THE DIVISION PRINCIPLE OF EQUALITY

The division principle is used when a number is multiplied times the unknown. The *division principle* states:

The same number can be divided into both sides of an equation without altering the equality.

The division principle is applied in solving for the unknown in the following equation:

$$10X = 100$$
$$\frac{10X}{10} = \frac{100}{10}$$
$$X = 10$$

In the example above, the unknown is multiplied times 10. The division principle applies. Ten is divided into both sides of the equation. On the left side of the equation, the 10's cancel each other out, isolating the unknown. On the right side, $100 \div 10 = 10$. Therefore, $X = 10$.

PRACTICE PROBLEM SET 7-4

Solve each equation.

1. $5X = 10$	2. $12Y = 47$	3. $19.2A = 63.5$
4. $0.146B = 71.05$	5. $1.16E = 22.08$	6. $44 = 19M$
7. $122 = 4.07N$	8. $39 = 3.13P$	9. $25 = 0.15Q$
10. $14.13 = 0.99R$		

THE ROOT AND POWER PRINCIPLES OF EQUALITY

The root principle is used when the unknown is raised to a power. The *root principle* states:

The same root can be taken of both sides of an equation without altering the equality.

The root principle is applied in solving for the unknown in the following equation:

$$X^2 = 16$$
$$\sqrt{X^2} = \sqrt{16}$$
$$X = 4$$

In the example above, the unknown is raised to the second power or squared. The root principle applies. The root is taken of both sides of the equation. On the left side of the equation, the square root of X^2 is X, isolating the unknown. On the right side, the square root of 16 is 4. Therefore, $X = 4$.

The power principle is used when a root is taken of the unknown. The *power principle* states:

Both sides of an equation can be raised to the same power without altering the equality.

The power principle is applied in solving for the unknown in the following equation:

$$\sqrt{X} = 10$$
$$(\sqrt{X})^2 = 10^2$$
$$X = 100$$

In the example above, the square root is taken of the unknown. The power principle applies. Both sides of the equation are raised to the second power or squared. On the left side of the equation, the square root of X is squared, isolating the unknown. On the right side, 10 squared equals 100. Therefore, $X = 100$.

PRACTICE PROBLEM SET 7-5

Solve each equation.

1. $X^2 = 9$
2. $Y^2 = 12$
3. $B^2 = 15.62$
4. $C^2 = 129.607$
5. $P^2 = 74.73$
6. $\sqrt{X} = 4$
7. $\sqrt{Y} = 3.2$
8. $\sqrt{B} = 5.56$
9. $\sqrt{C} = 7.62$
10. $\sqrt{P} = 9.07$

SOLVING COMPLEX EQUATIONS

The principles of equality just presented are tools that can be used to solve equations. All of the examples used to illustrate the principles thus far have been simple equations. However, most equations used in actual work settings are more complex than these examples.

Most equations dealt with on a daily basis consist of two or more algebraic operations:

$$10X + 7 = 52$$

The various principles of equality apply to these complex equations, too. However, there are several other rules that must be observed when solving complex equations. It is important that these rules be applied in the proper order. The rules follow and are listed in the proper sequence. When solving equations, all rules do not necessarily apply. If a rule does not apply to a given equation, skip it and move to the next step in sequence.

1. If the equation contains parentheses, remove them by performing the indicated operations.
2. Combine all like terms on both sides of the equation.
3. Arrange all unknown terms on one side of the equation and all known terms on the other side. This is accomplished by applying the addition and subtraction principles of equality.
4. Combine like terms where appropriate.
5. Apply the multiplication and division principles of equality where appropriate.
6. Apply the root and power principles of equality where appropriate.

The rules for solving complex equations are applied in solving the following sample problems:

SAMPLE 1

$$12x + 7 = 49 \quad \text{step 1}$$

$$12x + 7 - 7 = 49 - 7 \quad \text{step 2}$$

$$12x = 42 \quad \text{step 3}$$

$$\frac{12x}{12} = \frac{42}{12} \quad \text{step 4}$$

$$x = 3.5 \quad \text{step 5}$$

Step 1

The equation is examined to determine which rules apply. Rules 1, 2, 4, and 6 do not apply to this equation. Therefore, rules 3 and 5 are to be applied in order.

Step 2

The subtraction principle of equality is applied, allowing 7 to be subtracted from both sides of the equation.

Step 3

The 7's on the left side of the equation cancel each other out, leaving $12x$. On the right side $49 - 7 = 42$.

Step 4

The division principle of equality is applied, allowing both sides of the equation to be divided by 12.

Step 5

The unknown is isolated on the left side of the equation and 42 divided by 12 equals 3.5. Therefore, $x = 3.5$.

SAMPLE 2

$$5a + 4 = 3a - a + 20 \qquad \text{step 1}$$
$$5a + 4 = 2a + 20 \qquad \text{step 2}$$
$$5a + 4 - 4 = 2a + 20 - 4 \qquad \text{step 3}$$
$$5a = 2a + 16 \qquad \text{step 4}$$
$$5a - 2a = 2a - 2a + 16 \qquad \text{step 5}$$
$$3a = 16 \qquad \text{step 6}$$
$$\frac{3a}{3} = \frac{16}{3} \qquad \text{step 7}$$
$$a = 5.33 \qquad \text{step 8}$$

SAMPLE 3

$$12c + 5(c - 3) = 159.5 \qquad \text{step 1}$$
$$12c + 5c - 15 = 159.5 \qquad \text{step 2}$$
$$17c - 15 = 159.5 \qquad \text{step 3}$$
$$17c - 15 + 15 = 159.5 + 15 \qquad \text{step 4}$$
$$17c = 174.5 \qquad \text{step 5}$$
$$\frac{17c}{17} = \frac{174.5}{17} \qquad \text{step 6}$$
$$c = 10.26 \qquad \text{step 7}$$

SAMPLE 4

$$\frac{x^2 + 5}{2} = 39.72 \qquad \text{step 1}$$

$$2\left(\frac{x^2+5}{2}\right) = 2(39.72) \qquad \text{step 2}$$
$$x^2 + 5 = 79.44 \qquad \text{step 3}$$
$$x^2 + 5 - 5 = 79.44 - 5 \qquad \text{step 4}$$
$$x^2 = 74.44 \qquad \text{step 5}$$
$$\sqrt{x^2} = \sqrt{74.44} \qquad \text{step 6}$$
$$x = 8.63 \qquad \text{step 7}$$

SAMPLE 5

$$10\sqrt{x} = 43.41 \qquad \text{step 1}$$
$$\frac{10\sqrt{x}}{10} = \frac{43.41}{10} \qquad \text{step 2}$$
$$\sqrt{x} = 4.341 \qquad \text{step 3}$$
$$(\sqrt{x})^2 = (4.341)^2 \qquad \text{step 4}$$
$$x = 18.84 \qquad \text{step 5}$$

PRACTICE PROBLEM SET 7-6

Solve each equation.

1. $15a + 4 = 81$
2. $19.45p - 18.34 = 143.41$
3. $22x - 15 = 15x + 2x + 47$
4. $19c + 3c + c = 4c + 92.51$
5. $14d + 6(2d - 4) = 111.14$
6. $208 = 5x^2 + 2(x^2 - 5)$
7. $\dfrac{x^2 - 7}{3} = 45$
8. $\dfrac{x^2 + 9}{2} = 55.62$
9. $9\sqrt{p} = 9.4$
10. $15\sqrt{10r} = 15(\sqrt{4r} + 3)$

REVIEW PROBLEMS

Solve each equation.

1. $X - 12 = 36.17$
2. $52 = A \div 7.51$
3. $C - 8.85 = 41.33$
4. $a + 4.75 = 16.79$
5. $X + 0.019 = 6.6604$
6. $142.83 = R + 5.03$
7. $X/6 = 32.01$
8. $Y/3.35 = 24.12$
9. $10.64 = C/4.75$
10. $6X = 15$
11. $21.72R = 123.01$
12. $14.75 = 1.09X$
13. $X^2 = 17.51$
14. $Y^3 = 92.46$
15. $14a + 6 = 92$
16. $5X + 3X + 48 = 2.76X$
17. $18b + 1.56b + b = 136.51$
18. $12X + 5(3X - 6) = 108.56$
19. $10y^2 + 3(y^2 - 4) = 1{,}513$
20. $(5R + 4)(6R^2) = 99$

8

GEOMETRY: LINES AND ANGLES

Geometry is the branch of mathematics that deals with problems that focus on figures made of points and lines. An understanding of geometry is important to technicians.

The first step in learning geometry is to learn the definitions of several functional terms:

1. Point
2. Line
3. Parallel lines
4. Perpendicular lines
5. Oblique lines
6. Angles

DEFINITIONS

A point is the smallest geometric figure. Points represent locations in space; they have no length or depth. On paper a point is represented by a dot (.) or small characters consisting of intersecting lines (×) or (+). Points are usually given letter or number designations (i.e., point *A*, point *B*, point 1, point 2).

A line is the geometric figure created by connecting points through the shortest distance. The line connecting points *A* and *B* is called line *AB* (Figure 8-1). Theoretically a *line* extends infinitely. One that runs between two points is considered a *line segment.* However, for the purpose of sim-

Figure 8-1

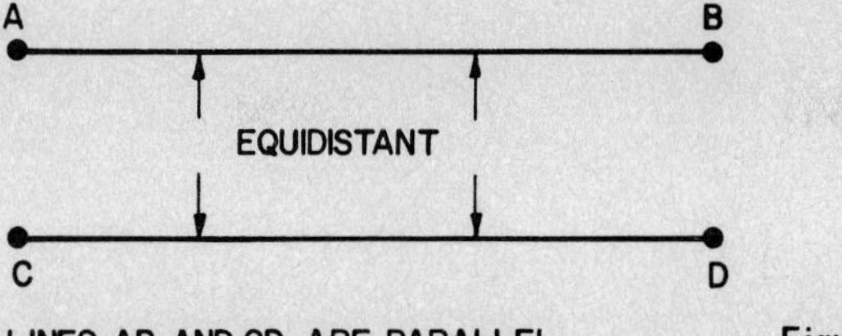

LINES AB AND CD ARE PARALLEL Figure 8-2

plicity, lines running between points will be referred to as lines. The term *line segment* will not be used in this book.

It should also be noted that there are straight lines and curved lines in geometry. However, again for simplicity, when the term *line* is used, it should be taken to mean straight line.

Parallel lines are lines that do not intersect. They are equidistant from each other (Figure 8-2). *Perpendicular lines* are lines that intersect at right angles to each other (Figure 8-3). *Oblique lines* are lines that intersect, but not at right angles (Figure 8-4).

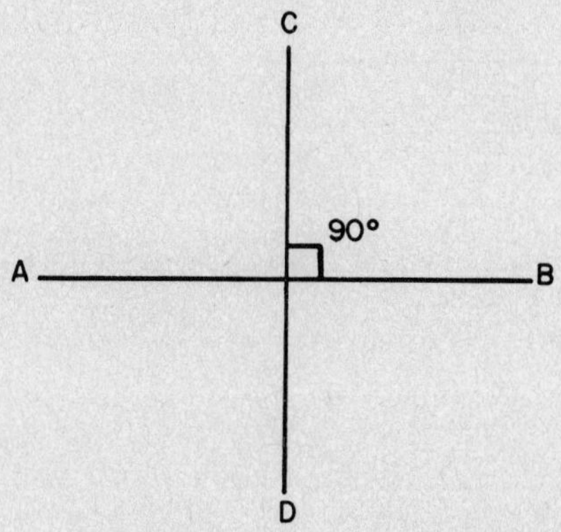

LINE AB AND CD ARE PERPENDICULAR Figure 8-3

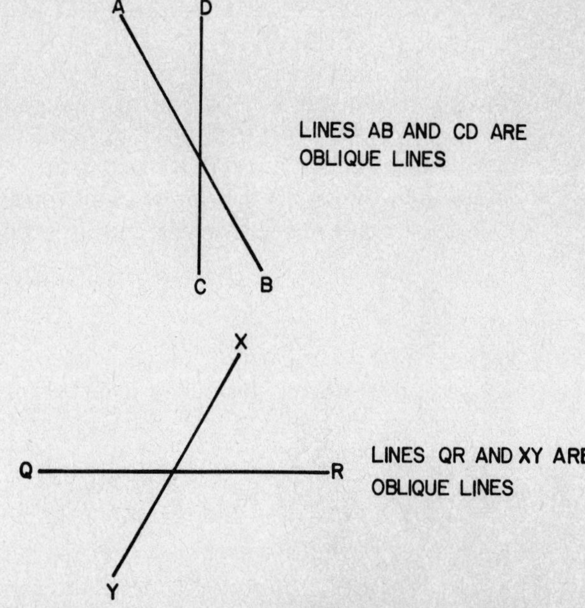

Figure 8-4

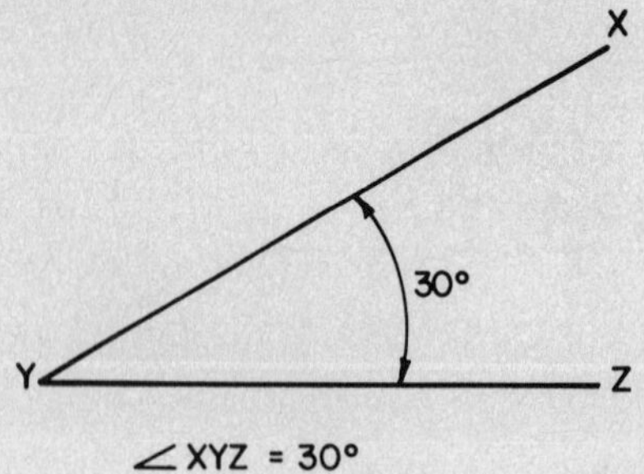

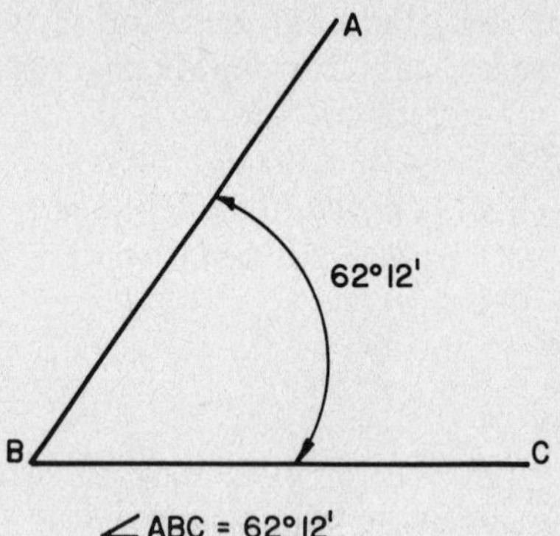

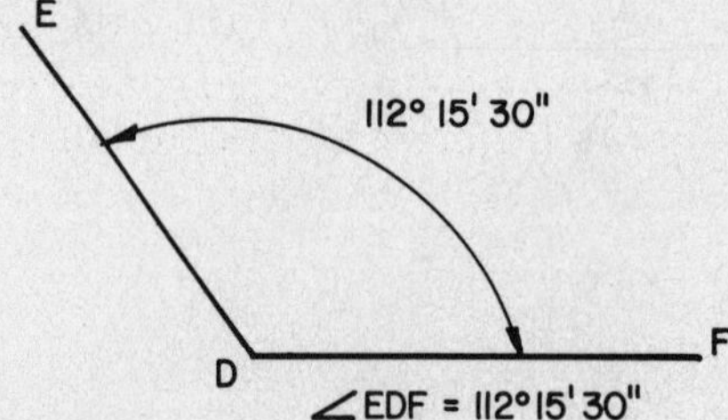

Figure 8-5

Lines that meet or intersect at a given point form *angles*. An angle has two *sides* formed by lines and a *vertex*, which is the point where the lines meet (Figure 8-5).

An angle is assigned a letter at its vertex and endpoints of the lines that form it. The symbol ($\angle$) and the assigned letters are used to call out an angle [i.e., $\angle XYZ$, $\angle ABC$, $\angle EDF$ (Figure 8-5)].

Angles are formed by using the vertex as a pivotal point and rotating one side or leg of the angle away from the other. The size of the resultant angle is measured in degrees, minutes, and seconds.

RULES OF GEOMETRY

The application of geometry is based on a set of rules. These rules are known as *axioms*. These axioms must be learned before attempting to solve problems of geometry.

1. Entities equal to the same thing or to other equal things are equal to each other.
2. Equals may be substituted for each other.

3. If equals are added to equals, the sums are equal.
4. If equals are subtracted from equals, the remainders are equal.
5. If equals are multiplied by equals, the products are equal.
6. If equals are divided by equals, the quotients are equal.
7. The whole of an entity is equal to the sum of its parts.
8. Only one straight line can be drawn between two points.
9. Only one line can be drawn parallel to another line through a given point.
10. Two straight lines can intersect at only one point.

ANGLES

There are five general classifications of angles:

1. Right angles
2. Acute angles
3. Obtuse angles
4. Straight angles
5. Reflex angles

A *right angle* is a 90° angle (Figure 8-6). An *acute angle* is less than 90° (Figure 8-6). An *obtuse angle* is greater than 90° but less than 180° (Figure 8-6). A *straight angle* is 180° or a straight line (Figure 8-6). A *reflex angle* is greater than 180° but less than 360° (Figure 8-6).

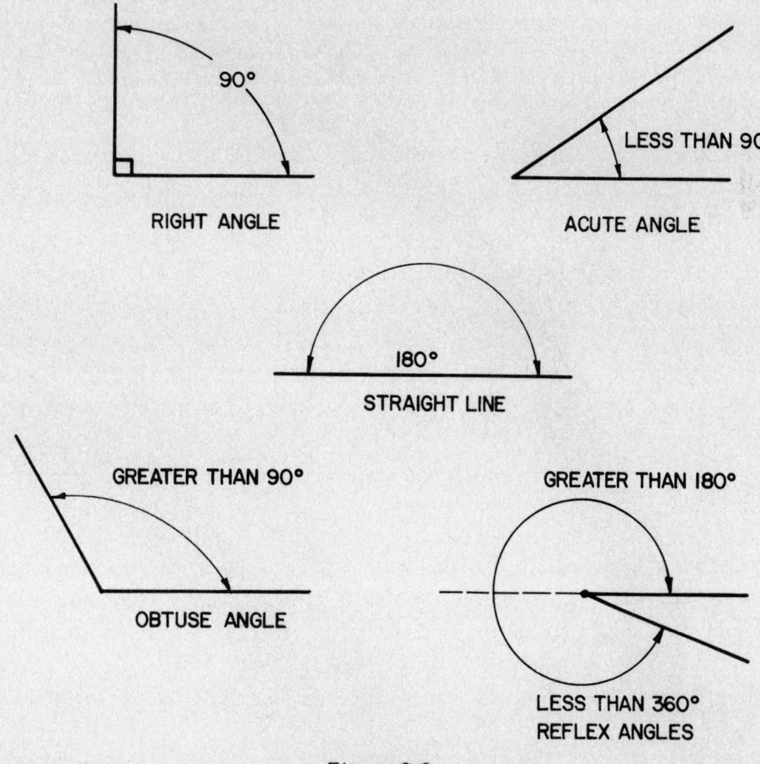

Figure 8-6

Geometric Rules for Angles

Earlier in this chapter a list of rules of geometry was presented. In this section another list of rules of geometry pertaining specifically to angles is presented.

Rule 1

When two lines intersect, opposite angles are equal (Figure 8-7).

Rule 2

When two parallel lines are intersected by a transversal, alternate interior angles are equal (Figure 8-7).

Rule 3

When two parallel lines are intersected by a transversal, corresponding angles are equal (Figure 8-8).

Rule 4

When two lines are intersected by a transversal and two corresponding angles are equal, the lines are parallel (Figure 8-8).

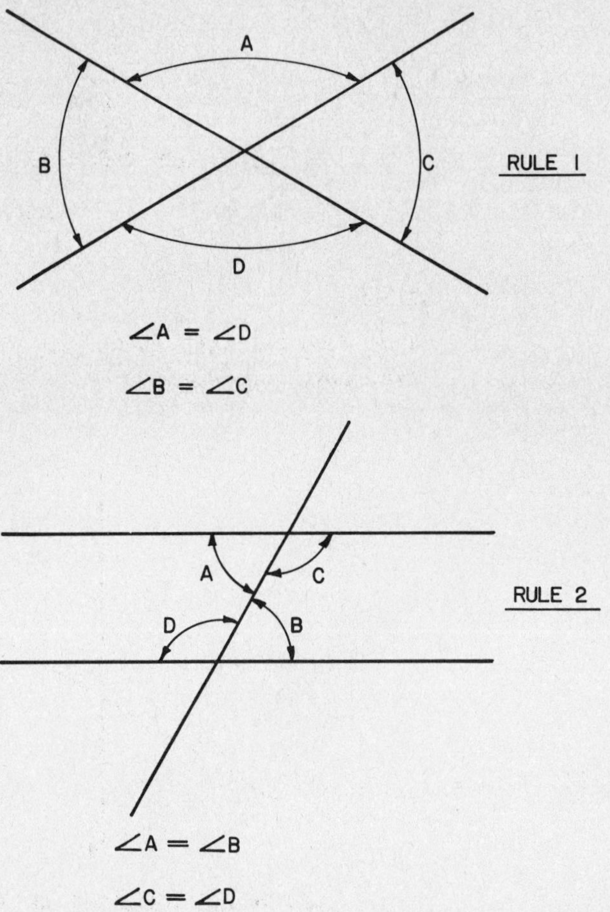

Figure 8-7

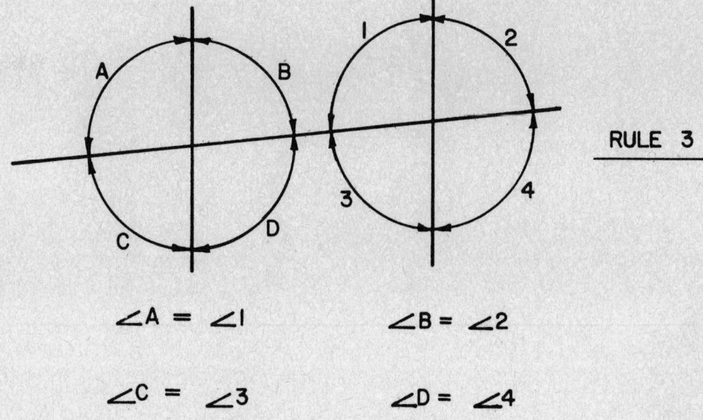

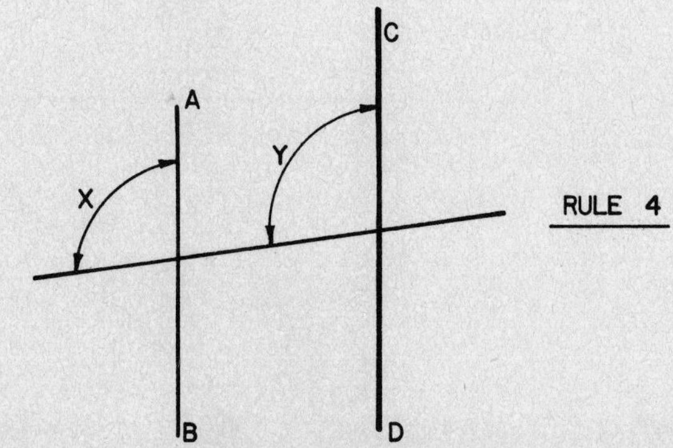

Figure 8-8

Rule 5

When two lines are intersected by a transversal and two alternate interior angles are equal, the lines are parallel (Figure 8-9).

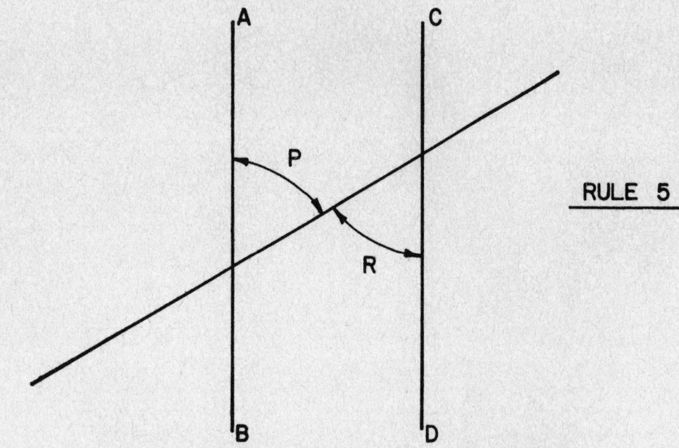

Figure 8-9

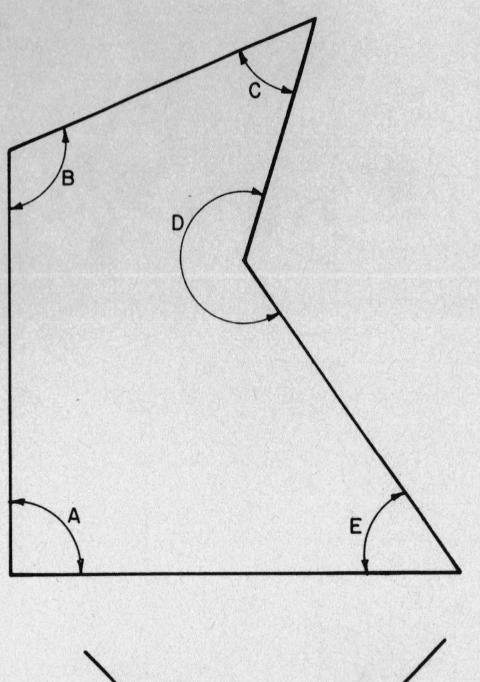

Figure 8-10

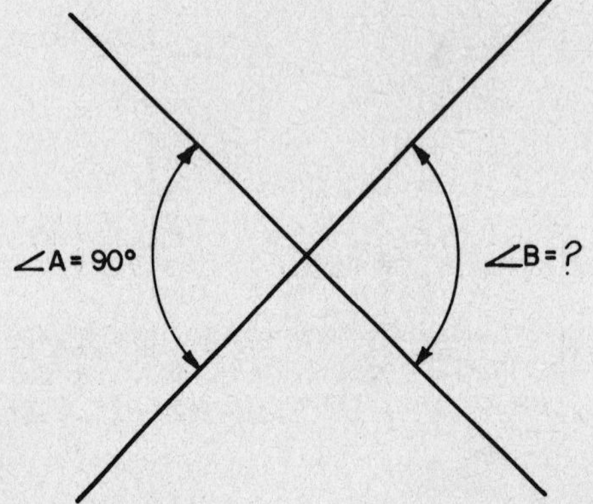

Figure 8-11

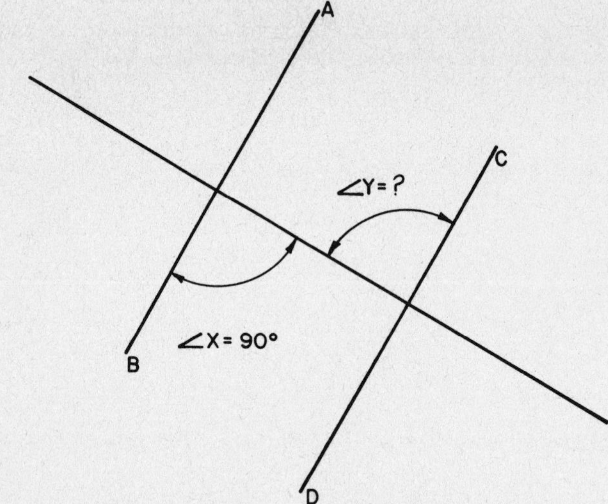

AB AND CD ARE PARALLEL

Figure 8-12

PRACTICE PROBLEM SET 8-1

Referring to Figure 8-10, identify each angle as acute, obtuse, straight, right, or reflex.

1. A = right
2. B = obtuse
3. C = acute
4. D = reflex
5. E = acute
6. In Figure 8-11, what is $\angle B$?
7. In Figure 8-12, what is $\angle Y$?
8. In Figure 8-13, are AB and CD parallel?
9. In Figure 8-14, if $\angle C = 77°$, what is $\angle B$?
10. In Figure 8-14, if $\angle A = 110°$, what is $\angle D$?

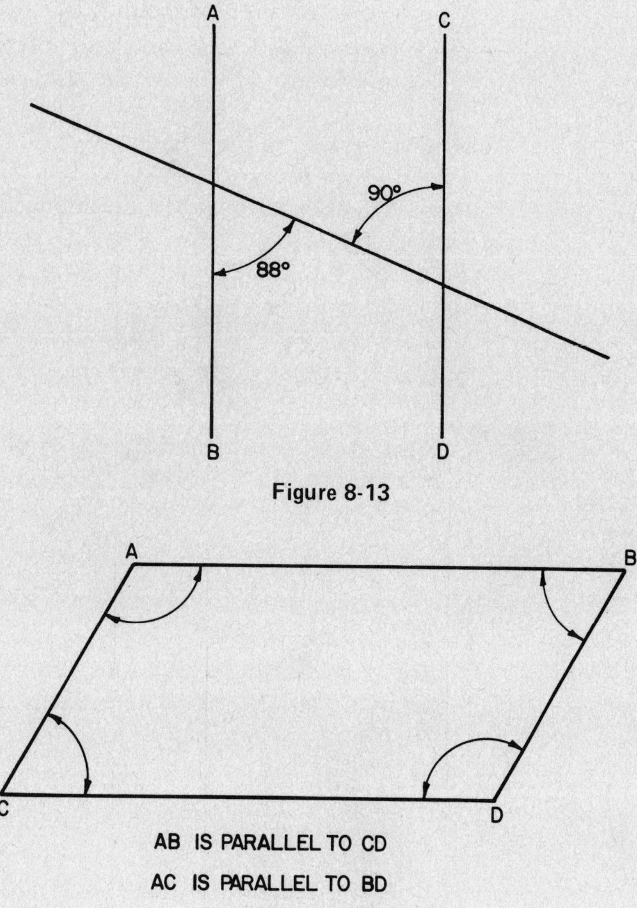

Figure 8-13

Figure 8-14

MEASURING ANGLES

Angles are measured in degrees, minutes, and seconds. There are 360 degrees in a circle, 60 minutes in a degree, and 60 seconds in a minute. The symbol for degrees is (°), the symbol for minutes is (′), and the symbol for seconds is (″).

Angles may be expressed in two ways: (1) as degrees, minutes, and seconds; and (2) as decimal degrees.

$$30°42'13''$$ (degrees–minutes–seconds form)

$$45.16°$$ (decimal form)

It is common practice to use the decimal form for adding and subtracting angles and to use the degrees–minutes–seconds form on drawings. In any case it is important for technicians to know:

1. How to convert the degrees–minutes–seconds form to the decimal form.
2. How to convert the decimal form to the degrees–minutes–seconds form.
3. How to add, subtract, multiply, and divide angles expressed in degrees, minutes, and seconds.
4. How to add, subtract, multiply, and divide angles expressed in decimal form.

Converting Angles to Decimal Form

It is much easier to perform mathematical operations on angles if they are expressed in decimal form. Angles expressed in degrees, minutes, and seconds can be converted to decimal form as follows:

$$14°23'12'' = 14.39°$$

Step 1

Convert 12″ to decimal form by dividing by 60 (there are 60 seconds in a minute.

$$\frac{12''}{60} = 0.2'$$

Step 2

Add the 0.2′ to the 23′ and convert the sum to decimal form by dividing by 60 (there are 60 minutes in a degree).

$$\frac{23.2'}{60} = 0.3866667°$$

Step 3

Round-off the 0.3866667° and add it to the 14°.

$$14.39°$$

PRACTICE PROBLEM SET 8-2

Convert each angle to decimal form.

1. 3°15′
2. 15°12′16″
3. 33°34′09″
4. 124°19′47″
5. 67°55′13″
6. 262°01′14″
7. 59°10′10″
8. 19°00′54″
9. 45°44′43″
10. 25°15′08″

Converting Decimal Angles to Degrees, Minutes, and Seconds

It is still common practice to state angles in the degress-minutes-seconds form on drawings. Angles expressed in decimal form can be converted to degrees, minutes, and seconds as follows:

$$19.62^\circ = 19^\circ 37' 12''$$

Step 1

Convert the 0.62° to minutes by multiplying by 60.

$$0.62 \times 60 = 37.2'$$

Step 2

Convert the $0.2'$ to seconds by multiplying by 60.

$$0.2 \times 60 = 12''$$

Step 3

Rewrite the decimal form as degrees, minutes, and seconds.

$$19^\circ 37' 12''$$

PRACTICE PROBLEM SET 8-3

Convert each decimal angle to degrees, minutes, and seconds.

1. 9.17°
2. 15.03°
3. 24.47°
4. 34.25°
5. 49.48°
6. 62.14°
7. 113.17°
8. 129.63°
9. 211.58°
10. 342.75°

Adding and Subtracting Degrees, Minutes, and Seconds

Degrees, minutes, and seconds can be added using normal addition procedures as follows:

$$\begin{array}{r} 18^\circ 39' 42'' \\ + \underline{25^\circ 44' 59''} \\ 44^\circ 24' 41'' \end{array}$$

Step 1

Begin at the extreme right by adding the seconds.

$$\begin{array}{r} 42'' \\ + \underline{59''} \\ 101'' \end{array}$$

When, as in this case, the sum is more than $60''$, 1 minute ($60''$) is carried over to the minutes column and the remaining seconds are entered under the seconds column.

$$\begin{array}{r} 42'' \\ + \underline{59''} \\ 41'' \end{array} \quad (101'' - 60'')$$

Step 2

Add the minutes column, including the 1 minute carried over from step 1.

$$\begin{array}{rl} 1' & \text{(carried from step 1)} \\ 39' & \\ +\ \underline{44'} & \\ 84' & \end{array}$$

When, as in this case, the sum is more than 60′, 1 degree (60′) is carried over to the degrees column and the remaining minutes are entered under the minutes column.

$$\begin{array}{rl} 1' & \\ 39' & \\ +\ \underline{44'} & \\ 24' & (84' - 60') \end{array}$$

Step 3

Add the degrees column, including the 1 degree carried over from step 2.

$$\begin{array}{r} 1^\circ \\ 18^\circ \\ +\ \underline{25^\circ} \\ 44^\circ \end{array}$$

Step 4

Rewrite the final sum.

$$44^\circ 24' 41''$$

Degrees, minutes, and seconds can be subtracted using normal subtraction procedures as follows:

$$\begin{array}{r} 24^\circ 12' 14'' \\ -\ \underline{18^\circ 15' 23''} \\ 5^\circ 56' 51'' \end{array}$$

Step 1

Begin at the extreme right by subtracting the seconds. When, as in this case, the larger number is being subtracted, you must borrow. Borrow 1 minute (60″) from the minutes column and perform the subtraction.

$$\begin{array}{rcr} 14'' + 60'' & = & 74'' \\ -\ \underline{23''} & & -\ \underline{23''} \\ & & 51'' \end{array}$$

Step 2

Subtract the minutes column, remembering that the top number has already been reduced by 1 minute. When, as in this case, the larger number is being subtracted, you must borrow. Borrow 1 degree (60′) from the degrees column and perform the subtraction.

$$\begin{array}{r} 11' + 60' = \\ -\underline{15'} \end{array} \quad \begin{array}{r} 71' \\ -\underline{15'} \\ 56' \end{array}$$

Step 3

Remember that the top number in the degrees column has been reduced by 1. Perform the subtraction operation.

$$\begin{array}{r} 23^\circ \\ -\underline{18^\circ} \\ 5^\circ \end{array}$$

Step 4

Rewrite the final result.

$$5^\circ 56' 51''$$

PRACTICE PROBLEM SET 8-4

Perform each addition or subtraction as indicated.

1. $\begin{array}{r} 16^\circ\ 14' \\ +\ \underline{12^\circ\ 15'} \end{array}$

2. $\begin{array}{r} 9^\circ\ 06'\ 17'' \\ +\ \underline{8^\circ\ 12'\ 19''} \end{array}$

3. $\begin{array}{r} 25^\circ\ 54'\ 58'' \\ +\ \underline{34^\circ\ 16'\ 42''} \end{array}$

4. $\begin{array}{r} 110^\circ\ 13'\ 42'' \\ +\ \underline{28^\circ\ 41'\ 33''} \end{array}$

5. $\begin{array}{r} 264^\circ\ 08'\ 09'' \\ +\ \underline{7^\circ\ 11'\ 22''} \end{array}$

6. $\begin{array}{r} 48^\circ\ 26'\ 54'' \\ -\ \underline{10^\circ\ 13'\ 19''} \end{array}$

7. $\begin{array}{r} 75^\circ\ 38'\ 44'' \\ -\ \underline{15^\circ\ 20'\ 22''} \end{array}$

8. $\begin{array}{r} 98^\circ\ 15'\ 15'' \\ -\ \underline{50^\circ\ 25'\ 25''} \end{array}$

9. $\begin{array}{r} 112^\circ\ 20'\ 25'' \\ -\ \underline{90^\circ\ 12'\ 30''} \end{array}$

10. $\begin{array}{r} 72^\circ\ 08'\ 09'' \\ -\ \underline{56^\circ\ 10'\ 10''} \end{array}$

Multiplying and Dividing Degrees, Minutes, and Seconds

You learned in Part 1 that multiplication is a shortcut for addition. Angles can be multiplied by applying normal multiplication procedures as follows:

$$\begin{array}{rrrr} & 13^\circ & 42' & 33'' \\ \times & & & 3 \\ \hline & 39^\circ & 126' & 99'' \\ & & +\,1 - & 60 \\ \hline & 39^\circ & 127' & 39'' \\ + & 2 - & 120 & \\ \hline & 41^\circ & 7' & 39'' \end{array}$$

Step 1

Beginning at the extreme right, multiply the seconds, minutes, and degrees times 3 and record the results under the appropriate column. This procedure yields a result of

$$39^\circ 126' 99''$$

Step 2

Because the seconds element is over 60, it must be reduced. Take as many minutes (multiples of 60) out of the seconds column as possible by subtracting. In this case, you can subtract 1 minute (60 seconds) from the seconds column and add it to the minutes column.

$$\begin{array}{rrr} 39^\circ & 126' & 99'' \\ & +1 & -60 \\ \hline 39^\circ & 127' & 33'' \end{array}$$

Step 3

Because the minutes element is over 60, it must be reduced. Take as many degrees (multiples of 60) out of the minutes column as possible by subtracting. In this case you can subtract 2 degrees (120 minutes) from the minutes column and add them to the degrees column to produce the final answer.

$$\begin{array}{rrr} 39^\circ & 127' & 39'' \\ +\ 2 & -120 & \\ \hline 41^\circ & 7' & 39'' \end{array}$$

You learned in Part 1 that division is a shortcut for subtraction. Angles can be divided by applying normal division procedures as follows:

$$\frac{43^\circ 13' 17''}{2} = 21^\circ 36' 39''$$

Step 1

Divide 43° by 2 and the result is 21° with a 1° remainder. The remainder is converted to minutes (60) and added to the minutes element.

$$\begin{array}{r} 21^\circ \\ 2\overline{)43^\circ} \\ \underline{42} \\ 1 \end{array} \quad (60 \text{ minutes}) + 13' = 73'$$

Step 2

Divide 73′ (13′ + 60′ remainder from step 1) by 2 and the result is 36′ with a 1′ remainder. The remainder is converted to seconds (60) and added to the seconds element.

$$\begin{array}{r} 36' \\ 2\overline{)73'} \\ \underline{72} \\ 1 \end{array} \quad (60 \text{ seconds}) + 17'' = 77''$$

Step 3

Divide 77″ (17″ + 60″ remainder from step 2) and the result is 38.5″. The final answer is

$$21^\circ 36' 39''$$

PRACTICE PROBLEM SET 8-5

Perform each multiplication or division as indicated.

1. $38^\circ\ 16' \times 2$
2. $49^\circ\ 39' \times 3$
3. $52^\circ\ 12'\ 09'' \times 4$
4. $16^\circ\ 10'\ 18'' \times 3.5$
5. $104^\circ\ 25'\ 14'' \times 5$
6. $76^\circ\ 32' \div 2$
7. $148^\circ\ 57' \div 3$
8. $208^\circ\ 48'\ 36'' \div 4$
9. $56^\circ\ 36'\ 03'' \div 3.5$
10. $522^\circ\ 06'\ 10'' \div 5$

REVIEW PROBLEMS

Convert each angle to decimal form.

1. $4^\circ16'$
2. $16^\circ13'18''$
3. $36^\circ36'12''$
4. $124^\circ18'37''$
5. $69^\circ56'15''$
6. $273^\circ09'16''$
7. $69^\circ11'12''$
8. $29^\circ00'54''$
9. $55^\circ55'44''$
10. $27^\circ17'09''$

Convert each decimal angle to degrees, minutes, and seconds.

11. 16.78°
12. 26.48°
13. 36.72°
14. 132.14°
15. 351.74°

Perform each addition or subtraction as indicated.

16. $\begin{array}{r} 17^\circ15' \\ +\ \underline{13^\circ14'} \end{array}$
17. $\begin{array}{r} 49^\circ27' \\ -\ \underline{12^\circ13'} \end{array}$
18. $\begin{array}{r} 09^\circ06' \\ +\ \underline{08^\circ11'} \end{array}$
19. $\begin{array}{r} 26^\circ56'48'' \\ -\ \underline{14^\circ58'49''} \end{array}$
20. $\begin{array}{r} 73^\circ09'08'' \\ +\ \underline{113^\circ15'19''} \end{array}$

9

GEOMETRY: TRIANGLES AND POLYGONS

Chapter 8 dealt with lines and angles. This chapter deals with geometric figures composed of lines and angles: triangles and polygons. A *triangle* is a closed, three-sided geometric figure. A *polygon* is a closed geometric figure of three *or more* sides. A triangle is a polygon.

TRIANGLES

Triangles are simple, three-sided polygons, composed of straight lines and angles. The symbol for triangle is $\triangle$. There are four types of triangles:

1. Right triangles
2. Equilateral triangles
3. Isosceles triangles
4. Scalene triangles

Right Triangles

A right triangle is any triangle that contains a *90° or right angle.* The longest side of a right triangle is the side opposite the right angle. This side is called the *hypotenuse* (Figure 9-1).

Equilateral Triangles

An equilateral triangle is any triangle with *three equal sides.* Correspondingly, the three angles of an equilateral triangle must also be equal (Figure 9-2).

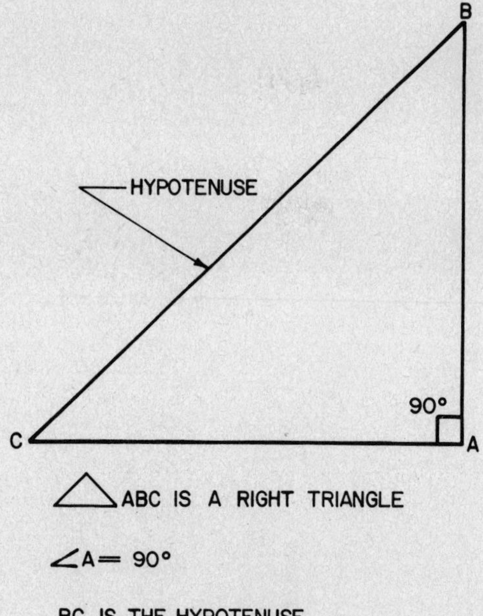

Figure 9-1

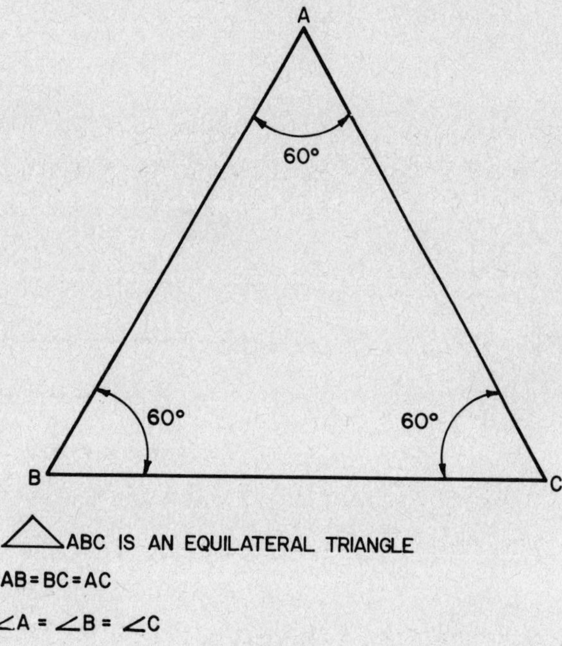

Figure 9-2

Isosceles Triangles

An isosceles triangle is any triangle with *two equal sides.* Correspondingly, it also has two equal angles (Figure 9-3).

Scalene Triangle

A scalene triangle is any triangle with *three unequal sides.* Correspondingly, the three angles of a scalene triangle are also unequal (Figure 9-4).

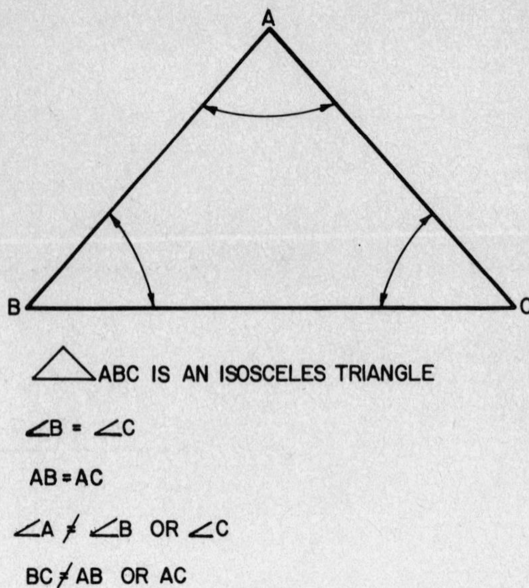

△ABC IS AN ISOSCELES TRIANGLE

∠B = ∠C

AB = AC

∠A ≠ ∠B OR ∠C

BC ≠ AB OR AC

Figure 9-3

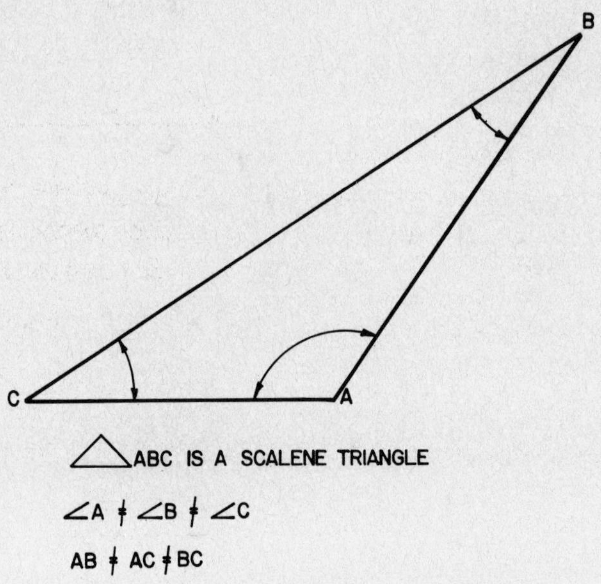

△ABC IS A SCALENE TRIANGLE

∠A ≠ ∠B ≠ ∠C

AB ≠ AC ≠ BC

Figure 9-4

SIDE-ANGLE RELATIONSHIPS OF TRIANGLES

There is a definite relationship that exists between sides and angles of triangles and it exists regardless of the type of triangle (right, equilateral, isosceles, or scalene).

Side Relationships

All triangles have three sides. Each side is opposite of an angle. That angle is known as the side's *corresponding angle* (Figure 9-5). The longest side of a triangle is always the side opposite the largest angle. The longer a side, the larger its corresponding angle.

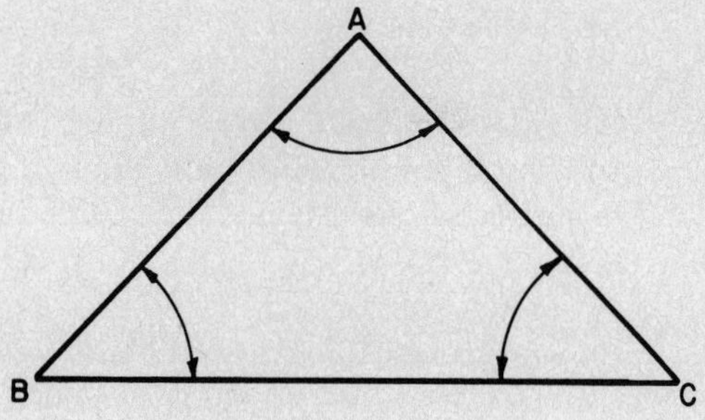

BC CORRESPONDS TO ∠A

AC CORRESPONDS TO ∠B

AB CORRESPONDS TO ∠C

Figure 9-5

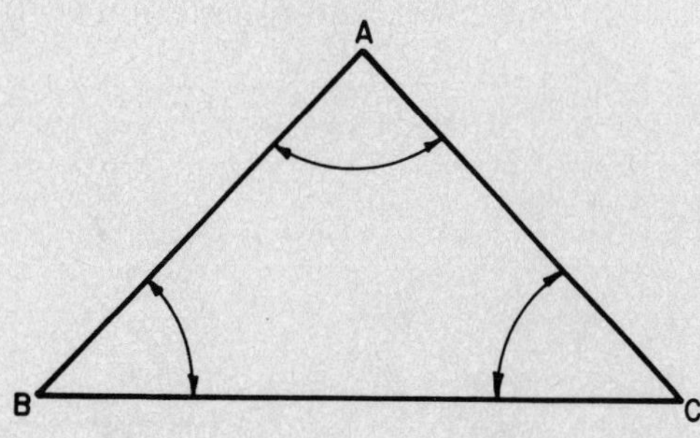

∠A CORRESPONDS TO BC

∠B CORRESPONDS TO AC

∠C CORRESPONDS TO AB

Figure 9-6

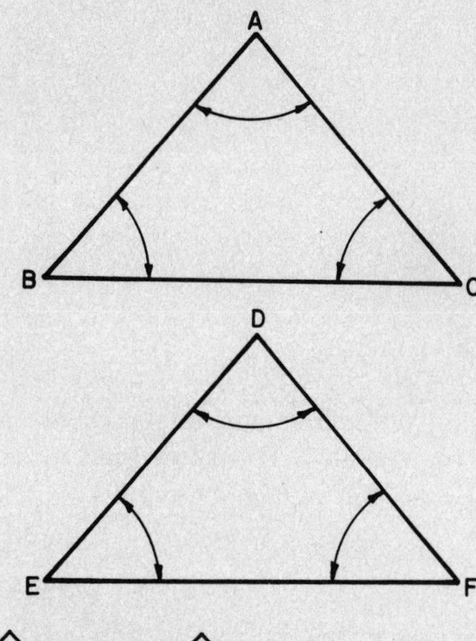

△ABC AND △DEF ARE CONGRUENT

AB = DE, AC = DF, BC = EF

∠A = ∠D , ∠B = ∠E , ∠C = ∠F

Figure 9-7

Angle Relationships

All triangles have three angles. Each angle is opposite a side. That side is the angle's corresponding side (Figure 9-6). The largest angle in a triangle is opposite the longest side. The larger an angle, the longer its corresponding side.

Congruency and Similarity Relationships

Relationships exist within triangles according to corresponding sides and angles. Relationships also exist *between* triangles. The relationships that can exist between triangles are *congruency* and *similarity* relationships.

Two *congruent triangles* are exact images of each other. This means that they are exactly the same in size, shape, and orientation (Figure 9-7).

Two *similar triangles* have equal corresponding angles but may have different sizes and orientations (Figure 9-8).

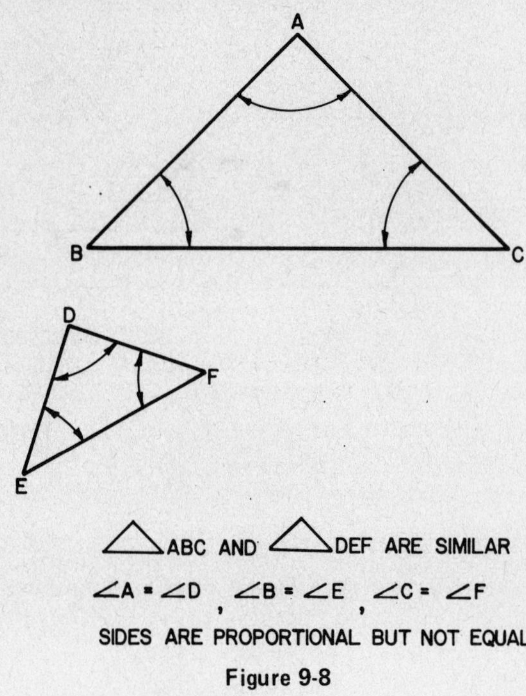

Figure 9-8

GEOMETRIC RULES FOR TRIANGLES

As with lines and angles, there are rules of geometry that apply specifically to triangles. The rules give technicians the tools they need for solving problems that involve triangles.

Rule 1

The sum of the angles contained in a triangle is always 180° (Figure 9-9).

Rule 2

Two triangles are similar when their corresponding sides are parallel (Figure 9-10).

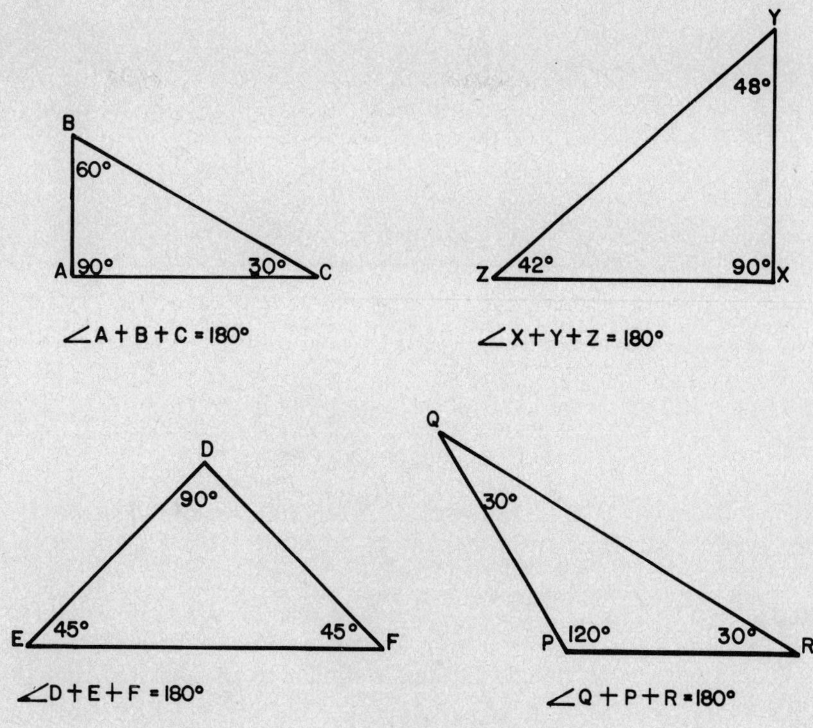

Figure 9-9

Rule 3

Two triangles are similar when their corresponding sides are perpendicular (Figure 9-11).

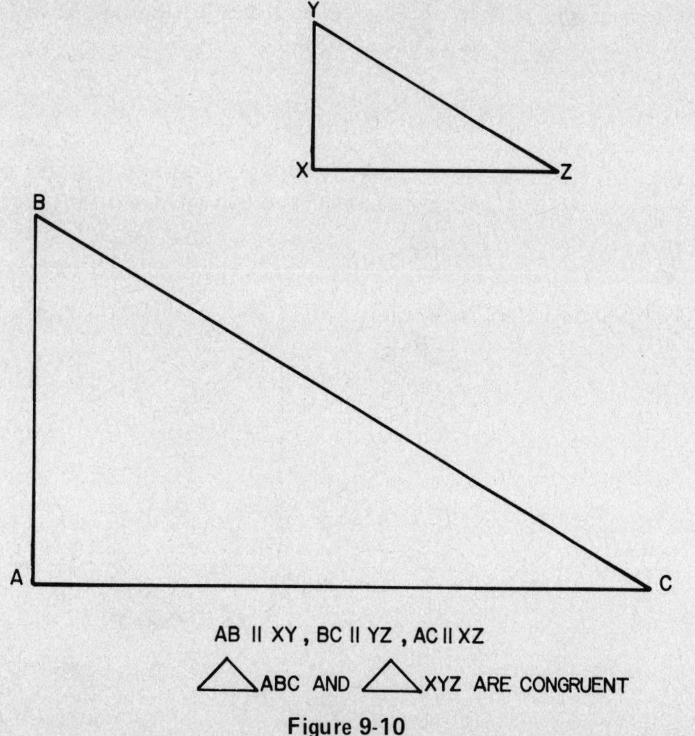

Figure 9-10

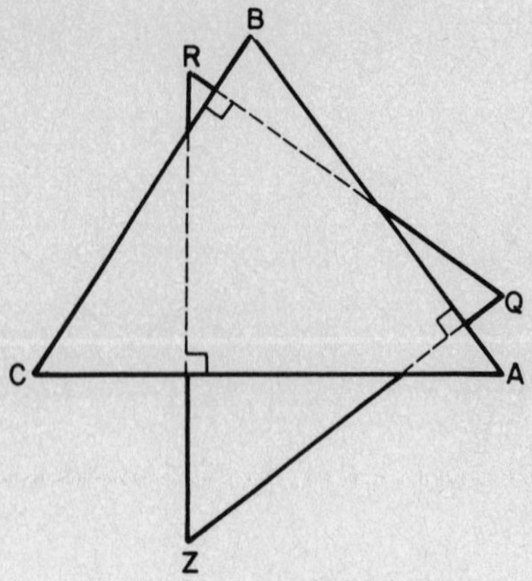

AB ⊥ QZ, BC ⊥ QR, AC ⊥ RZ

△ABC AND △QRZ ARE SIMILAR **Figure 9-11**

Rule 4

If a line is drawn inside of a triangle parallel to one side, the new triangle formed is similar to the original triangle (Figure 9-12).

Rule 5

In a right triangle, when a line is drawn from the vertex perpendicular to the corresponding side, the two new triangles formed are similar to the original triangle (Figure 9-13).

Rule 6

In an isosceles triangle an "altitude" bisects the base and the vertex angle (Figure 9-14).

Rule 7

In a right triangle, the square of the hypotenuse is equal to the sum of the squares of the other two sides (Figure 9-15).

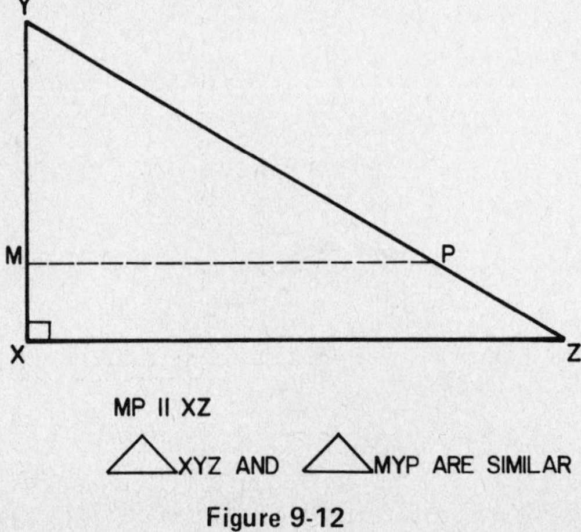

MP ∥ XZ

△XYZ AND △MYP ARE SIMILAR

Figure 9-12

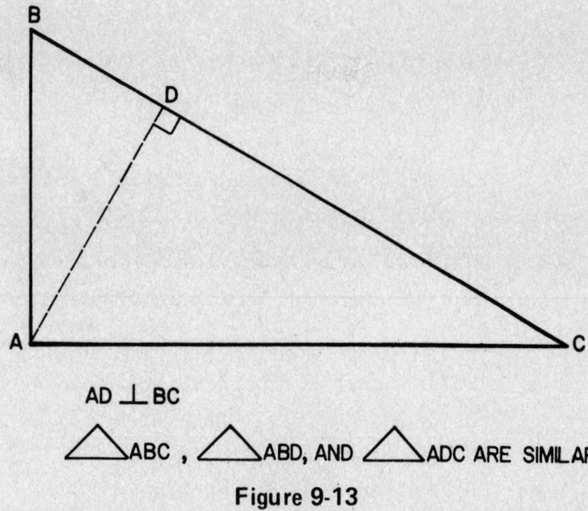

Figure 9-13

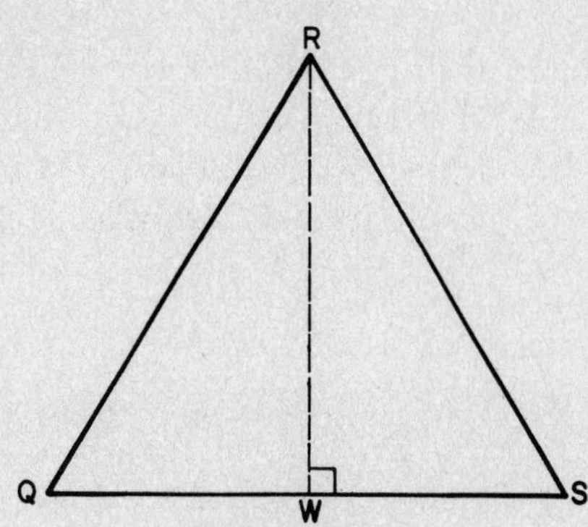

RW IS AN "ALTITUDE"
RW BISECTS QS
RW BISECTS ∠R

NOTE

AN ALTITUDE IS A LINE DRAWN FROM THE VERTEX PERPENDICULAR TO THE CORRESPONDING SIDE

Figure 9-14

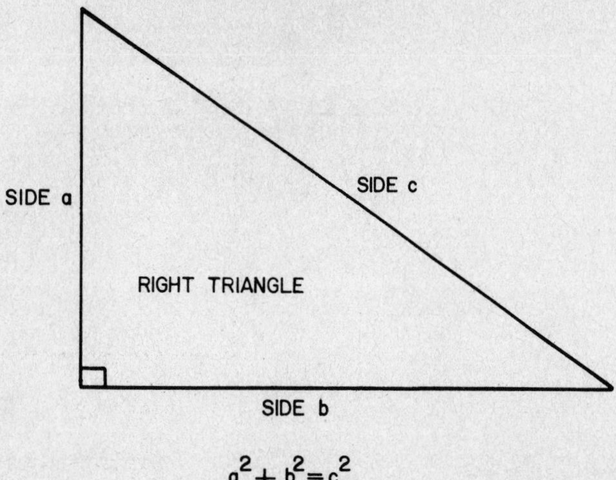

Figure 9-15

PRACTICE PROBLEM SET 9-1

Refer to Figure 9-16 to answer each question.

1. What type of triangle is ABC?
2. What type of triangle is XYZ?
3. What type of triangle is QRZ?
4. What type of triangle is MNP?
5. What type of angle is DEF?

Refer to Figure 9-17 to answer each question.

6. In triangle ABC, what is angle B?
7. In triangle XYZ, what is angle Z?
8. In triangle MNP, what is angle N?
9. In triangle DEF, what is angle E?
10. In triangle QRS, what is angle S?

Refer to Figure 9-18 to answer each question.

11. In triangle ABC, if side $a = 4$ and side $b = 6$, what is side c?
12. In triangle ABC, if side $a = 7$ and side $b = 9$, what is side c?
13. In triangle ABC, if side $c = 12$ and side $b = 10$, what is side a?
14. In triangle ABC, if side $c = 15$ and side $a = 9$, what is side b?

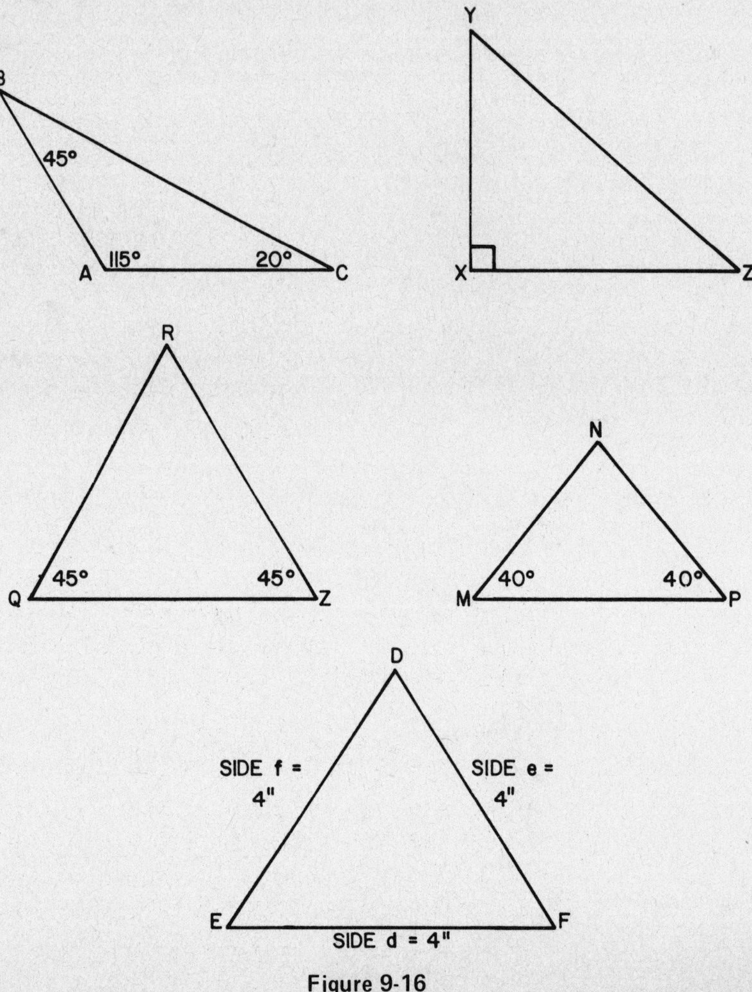

Figure 9-16

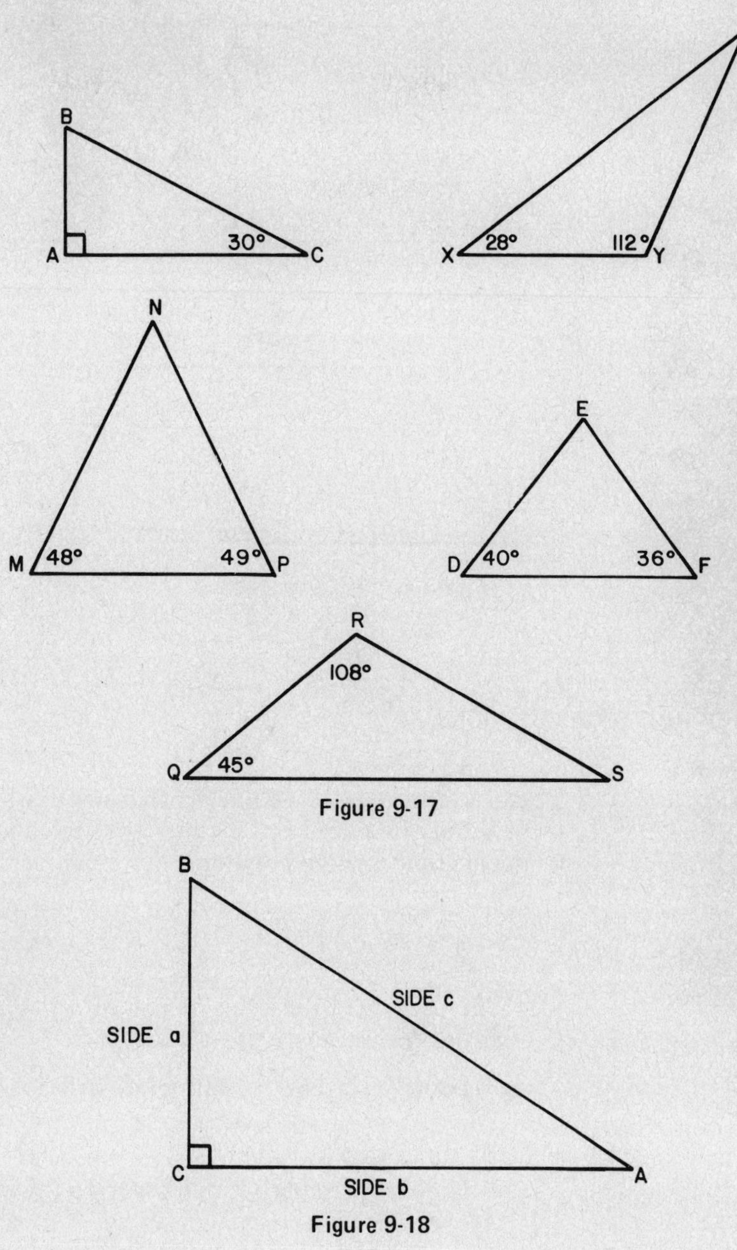

Figure 9-17

Figure 9-18

POLYGONS

Triangles are the simplest forms of polygons. There are several other geometric figures that are also polygons with which technicians should be familiar. These figures are illustrated in Figure 9-19.

1. A *square* is a four-sided figure in which all sides are equal and all four angles are 90° (Figure 9-19).
2. A *rectangle* is a four-sided figure in which opposite sides are equal and parallel and all angles are 90° (Figure 9-19).
3. A *parallelogram* is a four-sided figure in which opposite sides are equal and parallel, but the angles are not 90° angles (Figure 9-19).
4. A *regular hexagon* is a six-sided figure in which all sides and angles are equal (Figure 9-19).

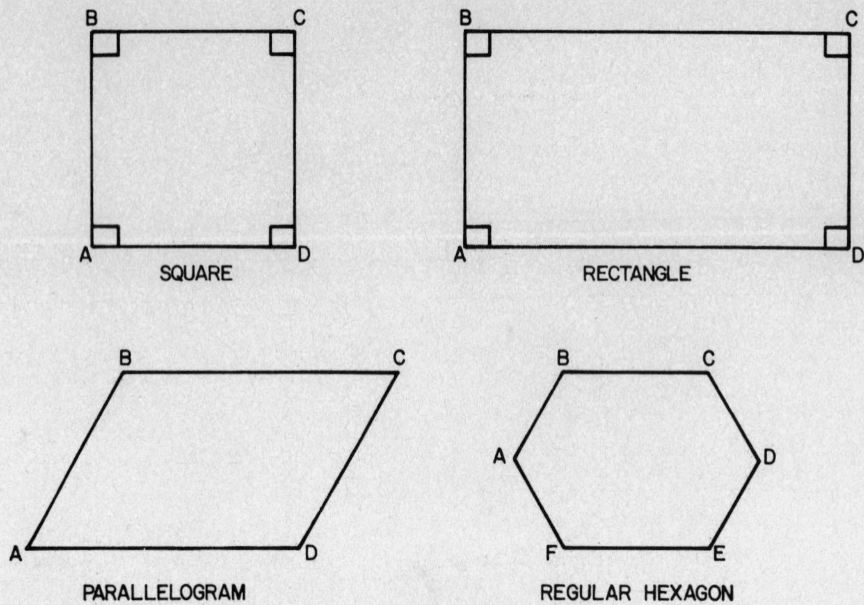

Figure 9-19

GEOMETRIC RULE FOR POLYGONS

There is one rule of geometry that applies to polygons with which technicians should be familiar. This rule gives technicians a valuable tool for solving problems that involve polygons.

Rule for Polygons

The sum of all interior angles of a polygon is equal to the number of angles (N) minus 2 times 180°.

$$(N-2)180° = \text{sum of all interior angles of a polygon}$$

PRACTICE PROBLEM SET 9-2

Refer to Figure 9-20 to solve each polygon for the angle indicated.

1. In polygon *ABCD*, solve for angle *A*.
2. In polygon *DFGH*, solve for angle *H*.
3. In polygon *JKLMNP*, solve for angle *J*.
4. In polygon *QRST*, solve for angle *R*.
5. In polygon *UVWXYZ*, solve for angle *Y*.

Refer to Figure 9-21 to solve each polygon for the angle indicated.

6. In polygon *ABC*, solve for angle *B*.
7. In polygon *DEFG*, solve for angle *F*.
8. In polygon *HJKL*, solve for angle *K*.
9. In polygon *MNPQR*, solve for angle *P*.
10. In polygon *STUVWX*, solve for angle *X*.

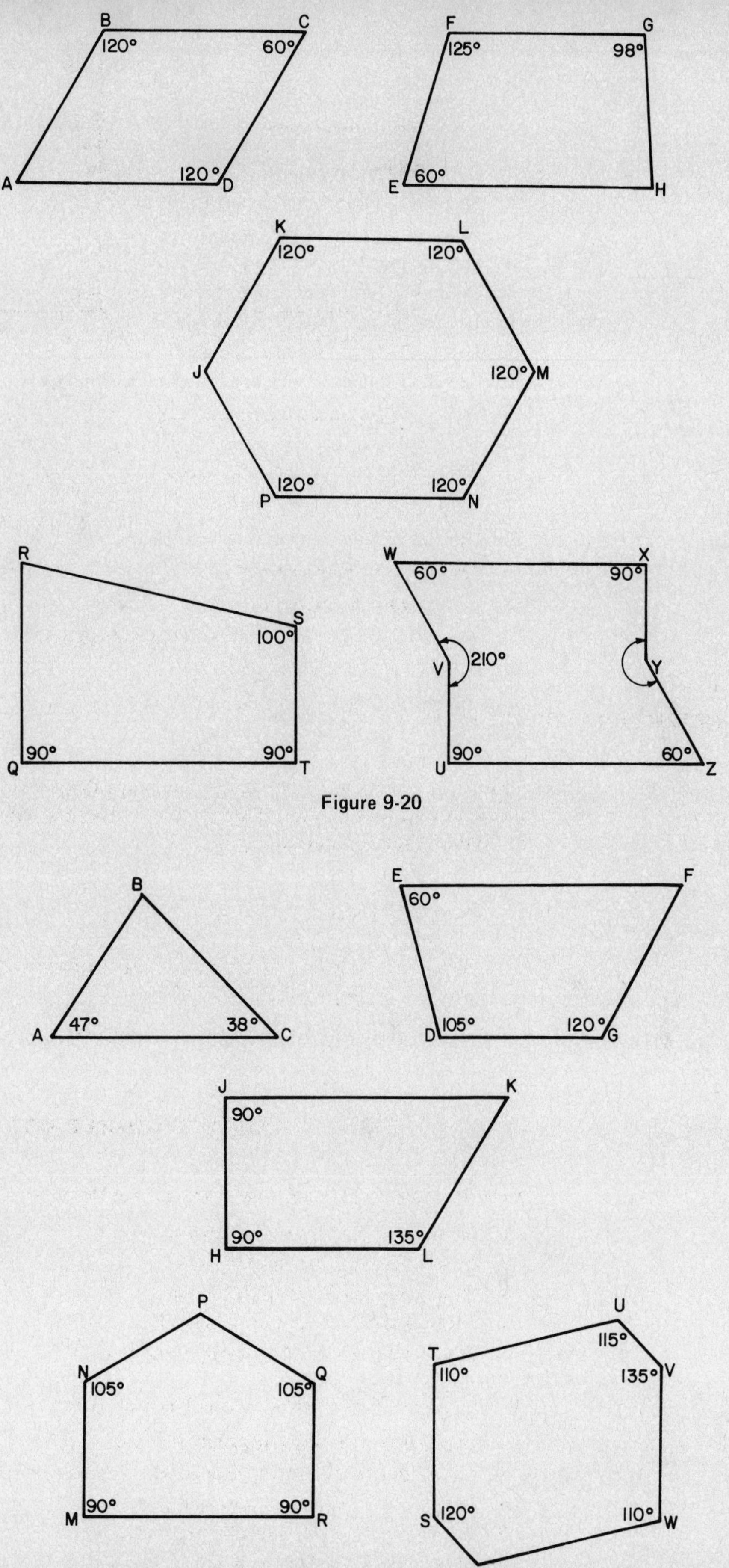

Figure 9-20

Figure 9-21

REVIEW PROBLEMS

1. Define the term "triangle."
2. Sketch the symbol for triangle.
3. What is the name of the longest side of a right triangle?
4. List the four types of triangles.
5. Define the term "equilateral triangle."
6. Define the term "isosceles triangle."
7. Define the term "scalene triangle."
8. Sketch examples of an isosceles, a scalene, and an equilateral triangle.
9. Define the term "corresponding angle."
10. Make a sketch that will illustrate the concept "corresponding angle."
11. Define the concept of "congruency."
12. Define the concept of "similarity."

Solve for the missing angle in each triangle.

13. $A = 12^\circ$, $B = 16^\circ$, $C = ?$
14. $A = 13.46^\circ$, $B = 72.01^\circ$, $C = ?$
15. $A = 121.16^\circ$, $B = 12.01^\circ$, $C = ?$
16. $A = 17.56^\circ$, $B = 36.06^\circ$, $C = ?$

Solve for the missing angle in each polygon.

17. $A = 14^\circ$, $B = 22^\circ$, $C = 37^\circ$, $D = ?$
18. $A = 17.51^\circ$, $B = 14.01^\circ$, $C = 48^\circ$, $D = 9^\circ$, $E = ?$
19. $A = 1.75^\circ$, $B = 88.52^\circ$, $C = ?$
20. $A = 15^\circ$, $B = 28.17^\circ$, $C = 52.02^\circ$, $D = ?$

10

GEOMETRY: CIRCLES AND ARCS

Two frequently used geometric figures are the circle and the arc. A *circle* is a closed curve in which all points are equidistant from the center point. An *arc* is a part of a circle between two specific points. The symbol for arc is ($\frown$). There are several definitions that apply to circles and arcs with which technicians should be familiar.

DEFINITIONS RELATING TO CIRCLES AND ARCS

To solve geometry problems involving arcs and circles, you will need to be familiar with the following definitions:

1. The *circumference* of a circle is the total length of the curved line that forms the circle (Figure 10-1).
2. A *chord* of a circle is any straight line joining two points on a circle (Figure 10-2).

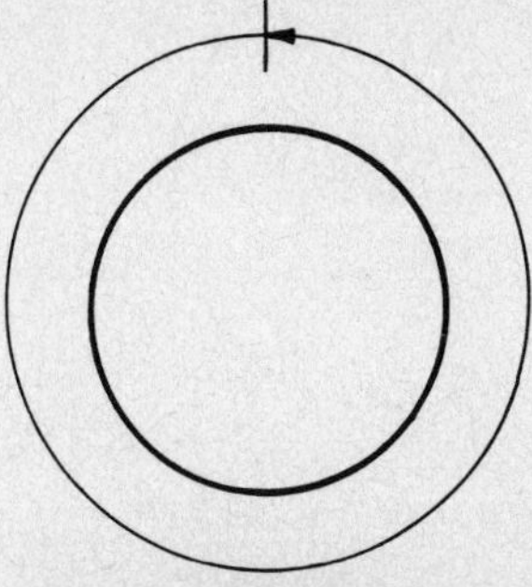

CIRCUMFERENCE Figure 10-1

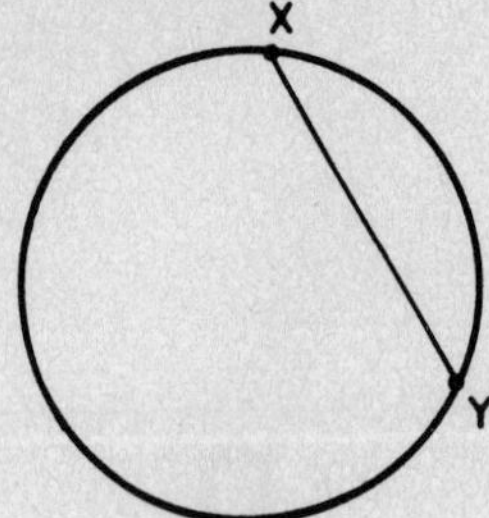

XY IS A CHORD Figure 10-2

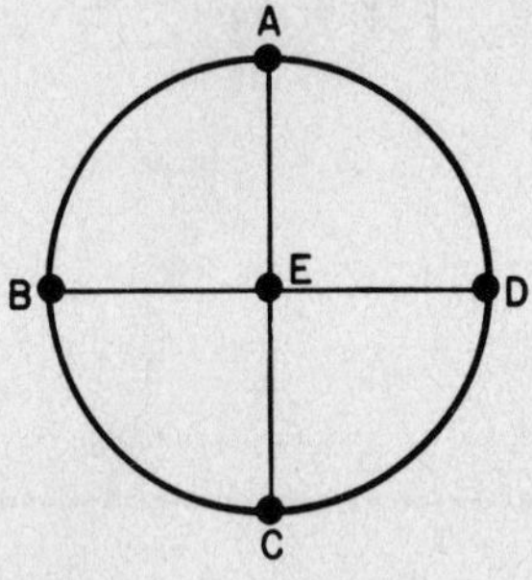

POINT E IS THE CENTER

AC = DIAMETER

BD = DIAMETER Figure 10-3

3. A *diameter* is a straight line or chord that intersects the center of a circle (Figure 10-3).
4. The *radius* of a circle is a straight line from the center to any point on the circle (Figure 10-4). (*Note:* The radius of a circle is one-half the diameter of the circle.)
5. A *tangent* of a circle is a straight line that touches but does not intersect the circle (Figure 10-5).
6. A *secant* of a circle is a straight line that passes through the circle and intersects it at two separate points (Figure 10-6).
7. A *segment* of a circle is an area marked off by a chord and its arc (Figure 10-7).

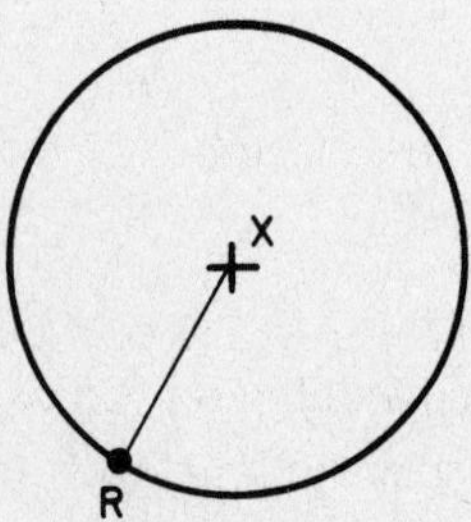

POINT X IS THE CENTER

XR = RADIUS Figure 10-4

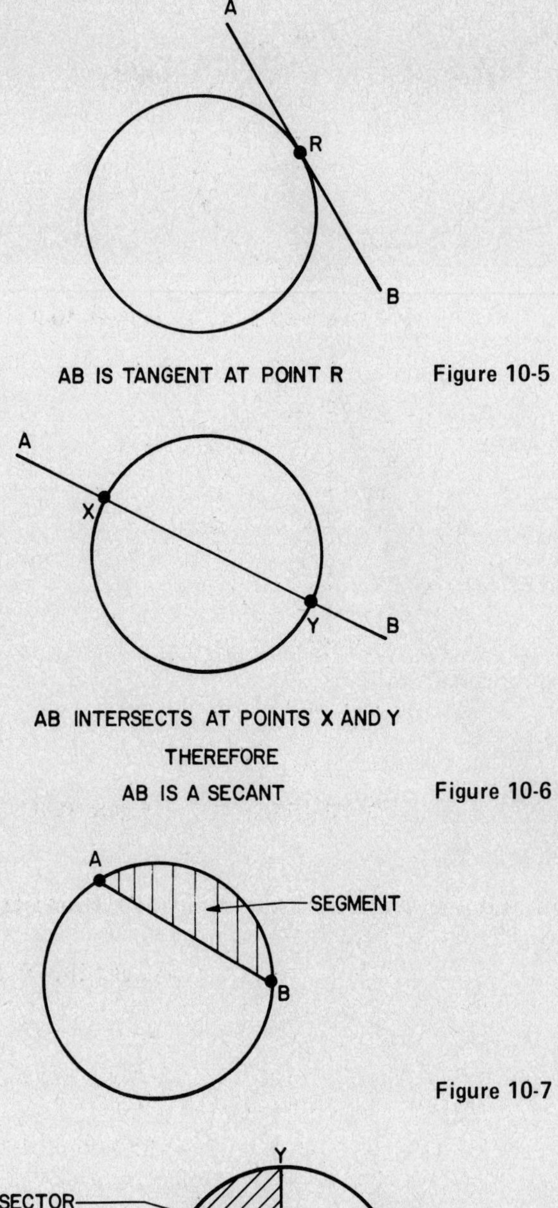

Figure 10-5

Figure 10-6

Figure 10-7

Figure 10-8

8. A *sector* of a circle is an area marked off by two radii and the corresponding arc (Figure 10-8).
9. A *central angle* of a circle is the angle formed by two radii (Figure 10-9).
10. An *inscribed angle* of a circle is any angle formed by chords in which the vertex lies *on* the circle (Figure 10-10).

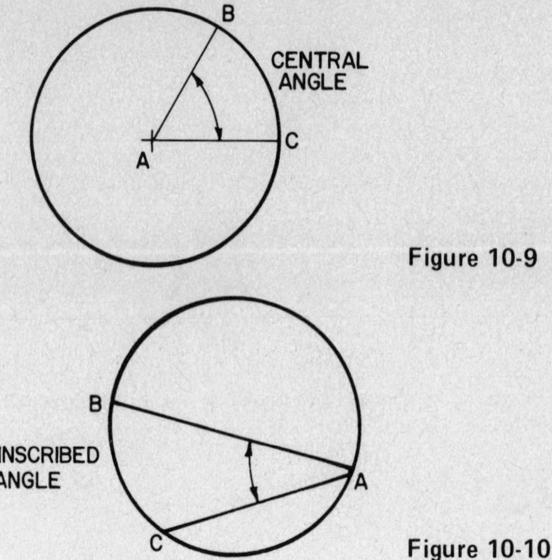

Figure 10-9

Figure 10-10

GEOMETRIC RULES FOR CIRCLES AND ARCS

There are several rules of geometry that apply to circles with which technicians should be familiar.

Rule 1

In the same circle, equal chords mark off equal arcs (Figure 10-11).

Rule 2

In the same circle, equal central angles mark off equal arcs (Figure 10-12).

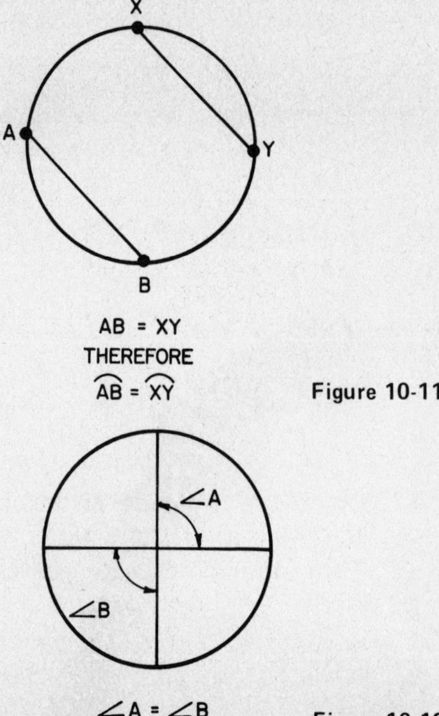

Figure 10-11

Figure 10-12

Rule 3

In the same circle, central angles have the same proportion as their corresponding arcs, and vice versa (Figure 10-13).

Rule 4

A tangent to a circle is perpendicular to a radius of the circle at the tangent point (Figure 10-14).

Rule 5

If two chords intersect inside a circle, the product of the two parts of one chord is equal to the product of the two parts of the other chord (Figure 10-15).

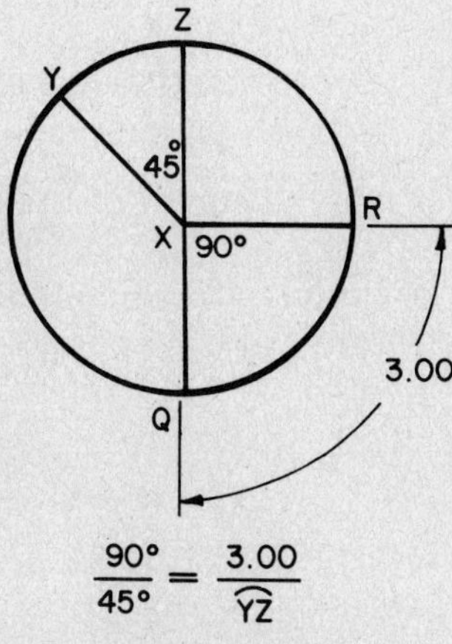

$$\frac{90°}{45°} = \frac{3.00}{\widehat{YZ}}$$

$$\widehat{YZ} = 1.5$$

Figure 10-13

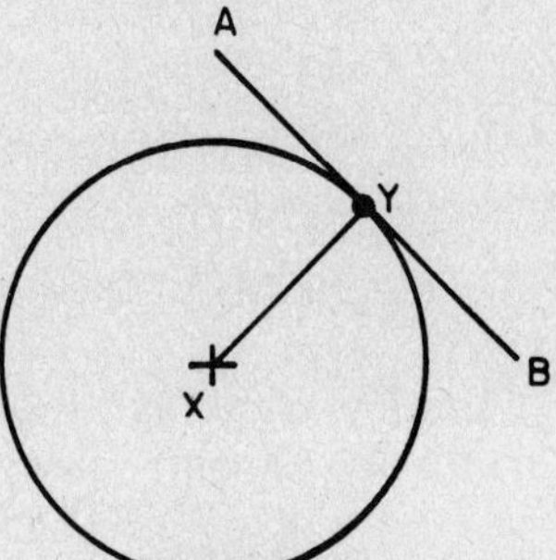

XY IS A RADIUS
AB IS TANGENT AT POINT Y
THEREFORE
$AB \perp XY$ AT POINT Y

Figure 10-14

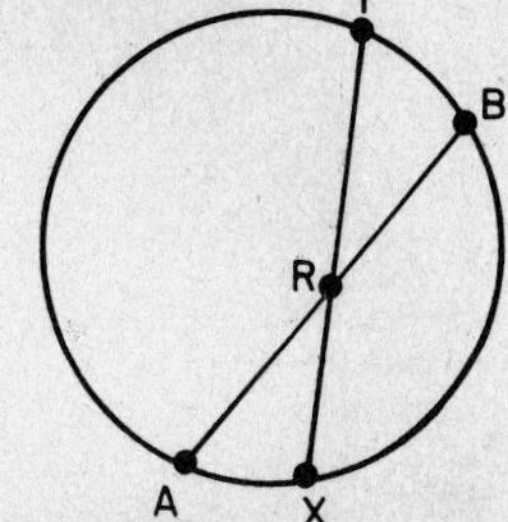

AB AND XY INTERSECT AT R
THEREFORE
(AR)(BR)=(XR)(YR)

Figure 10-15

FORMULAS FOR CIRCLES

There are several standard formulas used when working with circles with which technicians should be familiar. These formulas are presented and illustrated below.

Circumference Formulas

To find the circumference of a circle, multiply pi (π) times the diameter of the circle. (*Note:* $\pi = 3.1416$.)

$$C = \pi d$$

C = circumference of the circle

$\pi = 3.1416$

d = diameter of the circle

For example, if a circle has a diameter of 5 in., the circumference can be calculated using the formula above and substituting 5 for d.

$$C = \pi d$$
$$C = (3.1416)(5)$$
$$C = 15.71$$

Another version of the circumference formula uses the radius rather than the diameter of the circle:

$$C = 2r$$

C = circumference of the circle

$\pi = 3.1416$

r = radius of the circle

For example, if a circle has a radius of 3 in., the circumference can be calculated using the formula above and substituting 3 for r.

$$C = 2r$$
$$C = (2)(3.1416)(3)$$
$$C = 18.85$$

Diameter Formula

When the circumference of a circle is known but the diameter is not, the diameter can be calculated by dividing the circumference by pi (π).

$$D = \frac{C}{\pi}$$

D = diameter of the circle

C = circumference of the circle

$\pi = 3.1416$

For example, if a circle has a circumference of 17.86 in., the diameter can be calculated using the formula above and substituting 17.86 for C.

$$D = \frac{C}{\pi}$$

$$D = \frac{17.86}{3.1416}$$

$$D = 5.69$$

This same formula can be used for determining the radius of a circle by:

1. Solving for the diameter
2. Dividing the diameter by 2

Area Formula

If the radius or diameter of a circle is known, the area of the circle can be calculated using the following formula.

$$A = \pi r^2$$

$$r = \text{radius of the circle}$$

$$\pi = 3.1416$$

For example, if a circle has a radius of 6 in., the area can be calculated using the formula above and substituting 6 for r.

$$A = \pi r^2$$

$$A = (3.1416)(6^2)$$

$$A = (3.1416)(36)$$

$$A = 113.10$$

Another version of the formula may be used if the diameter is the known variable:

$$A = \frac{\pi D^2}{4}$$

$$A = \text{area of the circle}$$

$$D = \text{diameter of the circle}$$

$$\pi = 3.1416$$

For example, if a circle has a diameter of 12 in., the area can be calculated using the formula above and substituting 12 for D.

$$A = \frac{\pi D^2}{4}$$

$$A = \frac{(3.1416)(12^2)}{4}$$

$$A = \frac{(3.1416)(144)}{4}$$

$$A = 113.10$$

Another way of configuring the formula above is

$$A = 0.785D^2$$

This is a shortened version of the formula arrived at by dividing pi (π) by 4:

$$\frac{3.1416}{4} = 0.785$$

Of course, this version of the formula yields the same results, as can be seen in the following rework of the previous example:

$$A = 0.785D^2$$
$$A = (0.785)(12^2)$$
$$A = (0.785)(144)$$
$$A = 113.10$$

PRACTICE PROBLEM SET 10-1

Use the given information and the appropriate formula in each problem to solve for the unknown.

1. $D = 10$, $C = ?$
2. $D = 8$, $C = ?$
3. $D = 7.62$, $C = ?$
4. $r = 5.56$, $C = ?$
5. $r = 3.91$, $C = ?$
6. $C = 15.86$, $D = ?$
7. $C = 17.01$, $D = ?$
8. $r = 3.38$, $A = ?$
9. $r = 4.07$, $A = ?$
10. $D = 9.51$, $A = ?$

ARCS AND ANGLES IN AND OUT OF CIRCLES

There are several rules of geometry relating specifically to arcs, angles, and tangents of circles with which technicians should be familiar.

Rule 1

A central angle of a circle is equal to the angle of its corresponding intercepted arc (and is established by the radii that form the central angle) (Figure 10-16).

Rule 2

An inscribed angle of a circle is half of its corresponding intercepted arc (Figure 10-17).

Rule 3

The angle formed by a tangent and a chord that intersects the tangent point is half of its corresponding intercepted arc (Figure 10-18).

Rule 4

An angle formed by two secants and with the vertex outside a circle is equal to half the difference of its corresponding intercepted arcs (Figure 10-19).

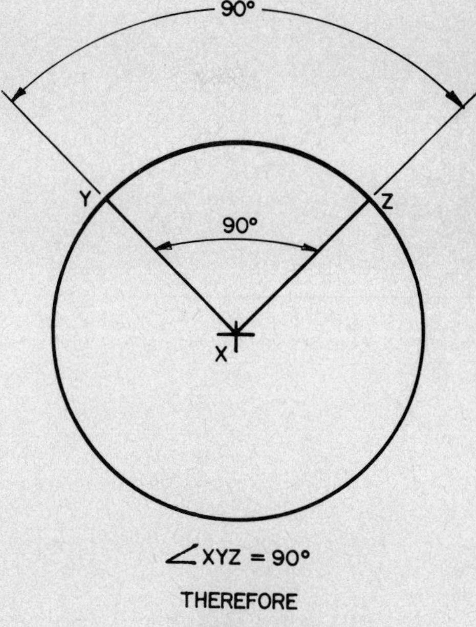

Figure 10-16

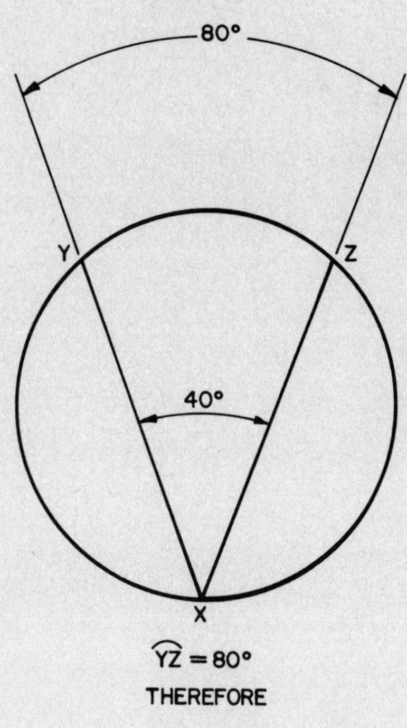

Figure 10-17

Rule 5

An angle formed by two tangents and with the vertex outside a circle is equal to half the difference of its corresponding intercepted arcs (Figure 10-20).

Rule 6

An angle formed by a tangent and a secant with the vertex outside a circle is equal to half the difference of its corresponding intercepted arcs (Figure 10-21).

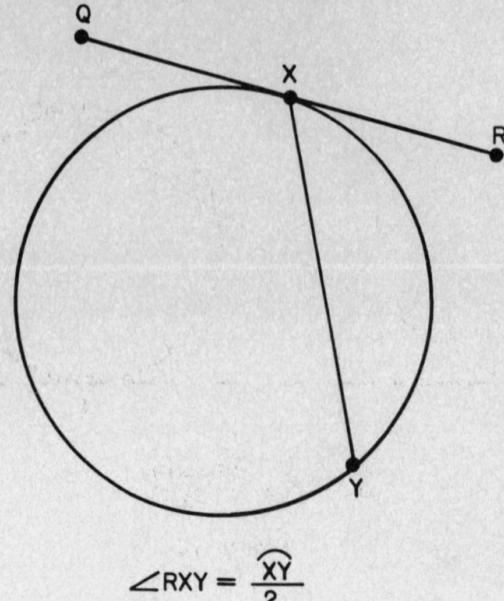

$\angle RXY = \frac{\widehat{XY}}{2}$

Figure 10-18

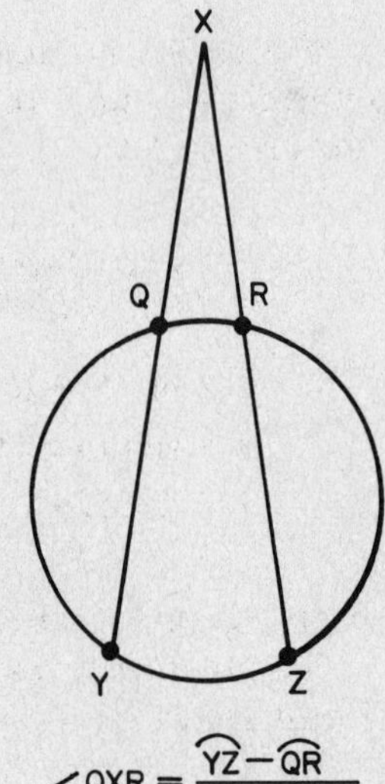

$\angle QXR = \frac{\widehat{YZ} - \widehat{QR}}{2}$

Figure 10-19

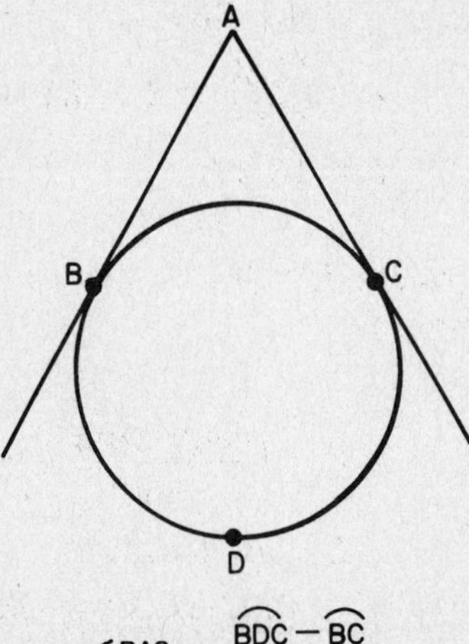

$\angle BAC = \frac{\widehat{BDC} - \widehat{BC}}{2}$

Figure 10-20

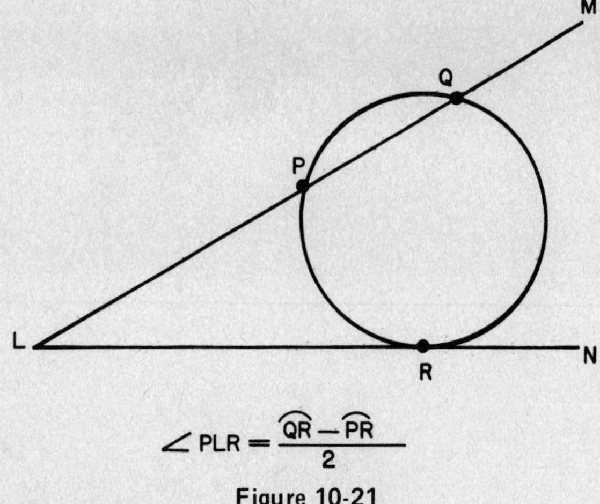

$$\angle PLR = \frac{\widehat{QR} - \widehat{PR}}{2}$$

Figure 10-21

FORMULAS FOR ARCS

Calculating the length of arcs is a common practice. There are two formulas that are helpful in determining the length of an arc. The first uses the arc degrees to determine the arc length:

$$L = \left(\frac{\text{arc degrees}}{360^\circ}\right)(2\pi r)$$

L = arc length

r = radius of the circle

$\pi = 3.1416$

The length of an arc can be calculated by substituting the arc degrees and radius of the circle into the formula above. For example, if the arc degrees are 120° and the radius is 5 in., the arc length can be calculated as follows:

$$L = \left(\frac{\text{arc degrees}}{360^\circ}\right)(2\pi r)$$

$$L = \left(\frac{120^\circ}{360}\right)(2 \times 3.1416 \times 5)$$

$$L = 10.47$$

The second formula for determining arc length uses the central angle and radius of the circle:

$$L = \left(\frac{\text{central angle}}{360^\circ}\right)(2\pi r)$$

L = arc length

r = radius of the circle

$\pi = 3.1416$

The length of an arc can be calculated by substituting the central angle and radius of the circle into the formula above. For example, if the central angle is 130° and the radius 6 in., the arc length can be calculated as follows:

$$L = \left(\frac{\text{central angle}}{360^\circ}\right)(2\pi r)$$

$$L = \left(\frac{130}{360}\right)(2 \times 3.1416 \times 6)$$

$$L = (0.36)(37.70)$$

$$L = 13.57$$

TANGENT CIRCLES

Just as a line may be tangent to a circle, circles may be tangent to each other. There are two tangency situations for circles:

1. Externally tangent circles
2. Internally tangent circles

Figure 10-22 illustrates both externally and internally tangent circles.

Externally Tangent Circles

There is a rule of geometry relating to externally tangent circles with which technicians should be familiar.

INTERNALLY TANGENT

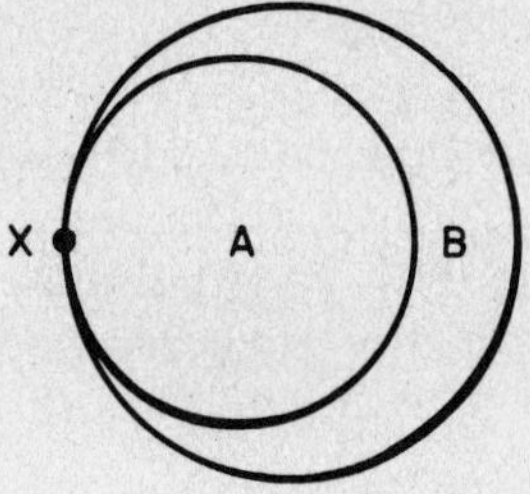

CIRCLES A AND B ARE TANGENT AT X

EXTERNALLY TANGENT

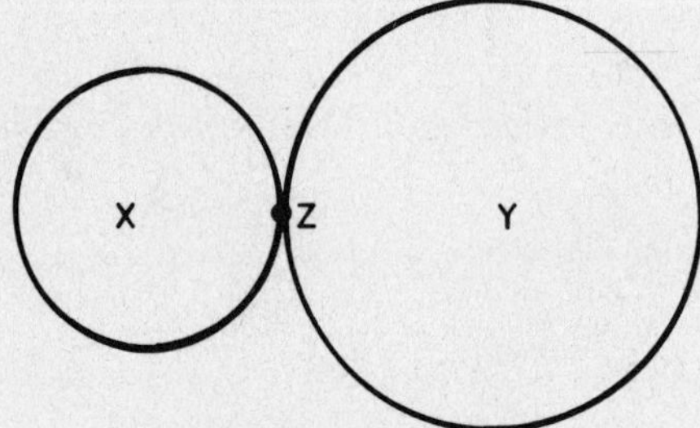

CIRCLES X AND Y ARE TANGENT AT Z

Figure 10-22

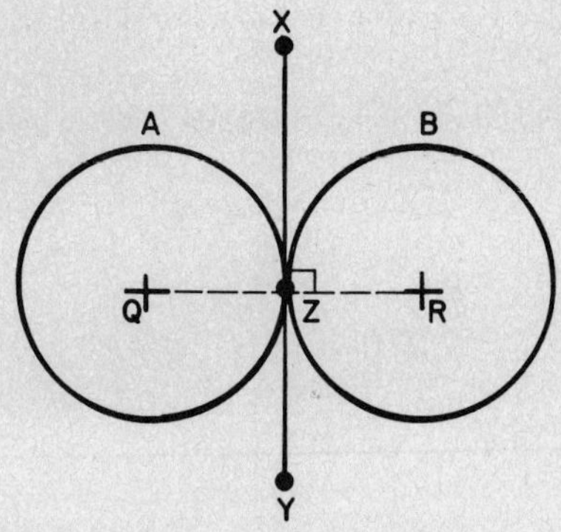

CIRCLES A AND B ARE TANGENT AT POINT Z

THEREFORE

QR ⊥ XY AND PASSES THROUGH POINT Z

Figure 10-23

Rule

If two circles are externally tangent, a line connecting the center points of the circles passes through the tangency point and is perpendicular to the tangent line (Figure 10-23).

Internally Tangent Circles

There is a rule of geometry relating to internally tangent circles with which technicians should be familiar.

Rule

If two circles are internally tangent, a line connecting the center points of the circles passes through the tangency point and is perpendicular to the tangent line (Figure 10-24).

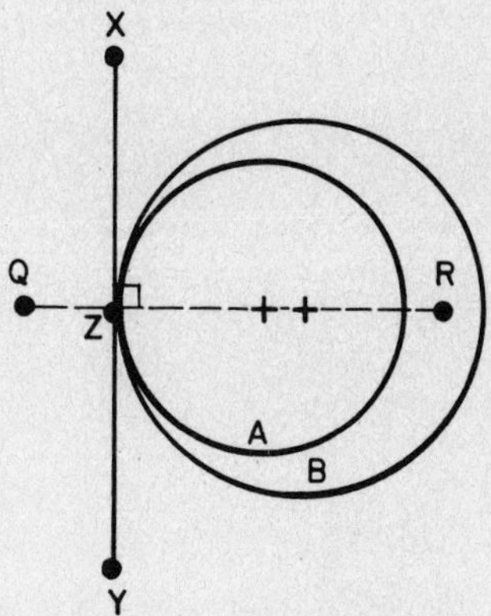

CIRCLES A AND B ARE TANGENT AT POINT Z

THEREFORE

QR ⊥ XY AND PASSES THROUGH POINT Z

Figure 10-24

PRACTICE PROBLEM SET 10-2

Given the information in each problem, calculate the arc length.

1. Arc degrees = 97°, $r = 5$, $L = ?$
2. Arc degrees = 15°, $r = 6$, $L = ?$
3. Arc degrees = 38°, $r = 9$, $L = ?$
4. Arc degrees = 47°, $r = 10.01$, $L = ?$
5. Arc degrees = 123°, $r = 7.53$, $L = ?$
6. Central angle = 107°, $r = 7$, $L = ?$
7. Central angle = 17°, $r = 3$, $L = ?$
8. Central angle = 39°, $r = 4$, $L = ?$
9. Central angle = 51°, $r = 15.01$, $L = ?$
10. Central angle = 125°, $r = 12.38$, $L = ?$

REVIEW PROBLEMS

Use the given information to solve for the unknown in each problem (D = diameter; C = circumference; r = radius; A = area).

1. $D = 12$, $C = ?$
2. $D = 7$, $C = ?$
3. $D = 5.56$, $C = ?$
4. $r = 7.69$, $C = ?$
5. $r = 4.98$, $C = ?$
6. $C = 16.68$, $D = ?$
7. $C = 19.03$, $D = ?$
8. $r = 6.38$, $A = ?$
9. $r = 4.09$, $A = ?$
10. $D = 19.51$, $A = ?$

Use the given information to calculate the arc length in each problem.

11. Arc degrees = 107°, $r = 7$
12. Arc degrees = 16°, $r = 6$
13. Arc degrees = 48°, $r = 12$
14. Arc degrees = 57°, $r = 10.15$
15. Arc degrees = 133°, $r = 8.56$
16. Central angle = 108°, $r = 9$
17. Central angle = 19°, $r = 5$
18. Central angle = 49°, $r = 8$
19. Central angle = 59°, $r = 16.16$
20. Central angle = 145°, $r = 13.83$

11
TRIGONOMETRY: RIGHT TRIANGLES

The right triangle is a common geometric shape. Therefore, it is important for technicians to know how to solve right triangles. Solving a right triangle means using its known elements and various trigonometric concepts to determine the unknown elements (e.g., length of sides or sizes of angles).

You learned earlier that a right triangle is any triangle that contains a 90° angle. Figure 11-1 illustrates an accepted method for labeling the various elements of a right triangle. The angles are labeled with capital letters (A, B, and C) and their corresponding sides are labeled with the corresponding lowercase letters (e.g., side a is opposite angle A).

Any combination of letters may be used, but the relationship of corresponding upper- and lowercase letters should be maintained. An angle and its corresponding side should be labeled with upper- and lowercase versions of the same letter.

There are two methods used for solving right triangles:

1. Pythagorean theorem
2. Trigonometric functions

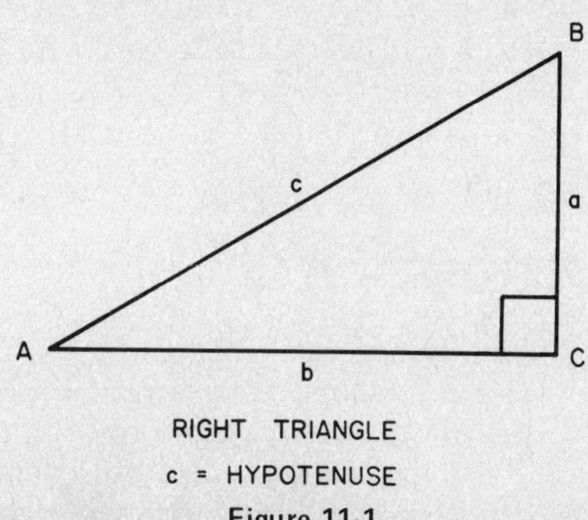

RIGHT TRIANGLE
c = HYPOTENUSE
Figure 11-1

PYTHAGOREAN THEOREM

This theorem states that in a right angle, the square of the hypotenuse is equal to the sum of the squares of the other two sides. By using the standard format for a right triangle shown in Figure 11-1, this theorem can be reduced to a formula:

$$a^2 + b^2 = c^2$$

This formula can be used for solving for the missing side of a right triangle when two sides are known (see Figure 11-2). In Figure 11-2, side $a = 5$ and side $b = 8$. By substituting these values into the formula above, the length of side c can be determined.

$$5^2 + 8^2 = c^2$$
$$25 + 64 = c^2$$
$$89 = c^2$$
$$9.43 = c$$

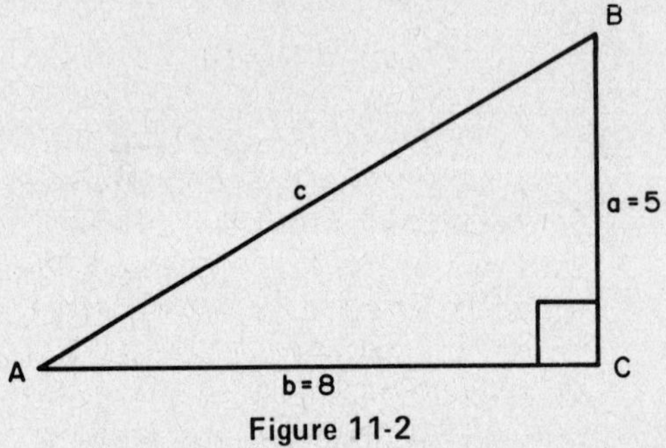

Figure 11-2

By applying the Pythagorean theorem to the solution of the right triangle in Figure 11-2, it was determined that side $c = 9.43$.

PRACTICE PROBLEM SET 11-1

In each problem, use the given sides and the Pythagorean theorem to solve for the missing side.

1. $a = 6$, $b = 8$, $c = ?$
2. $a = 8$, $b = 10$, $c = ?$
3. $a = 9$, $b = 11$, $c = ?$
4. $a = 10$, $b = 12$, $c = ?$
5. $a = 12.5$, $b = 14.5$, $c = ?$
6. $a = ?$, $b = 9$, $c = 12$
7. $a = ?$, $b = 7$, $c = 15$
8. $a = 10.58$, $b = ?$, $c = 17.63$
9. $a = 4.03$, $b = ?$, $c = 7.11$
10. $a = ?$, $b = 6.12$, $c = 18.05$

TRIGONOMETRIC FUNCTIONS

The Pythagorean theorem is applicable only when solving for a missing side and can be used only when two sides are known. Two known sides represent one of the two cases when a right triangle can be solved. The other case is when one side and one acute angle are known. In their case, right triangles can be solved by applying the trigonometric functions of an angle:

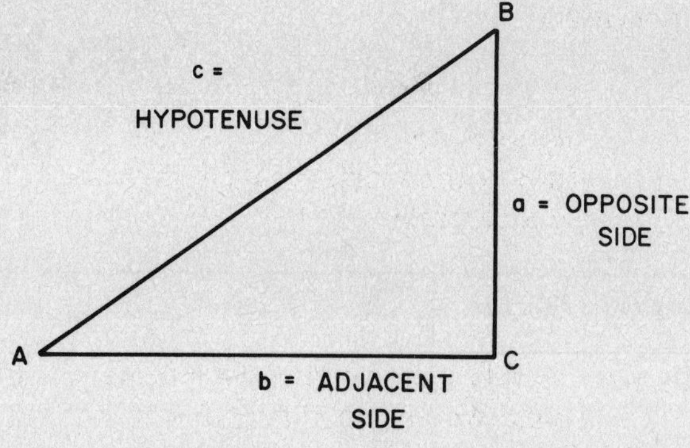

Figure 11-3

1. Sine
2. Cosine
3. Tangent
4. Cotangent
5. Secant
6. Cosecant

Of the six trigonometric functions listed above, the first three are the ones most frequently used for solving right triangles. To begin to understand trigonometric functions, examine the right triangle in Figure 11-3.

Each angle in a triangle has six possible trigonometric functions (see the list below). In solving right triangles, it is common practice to name angle A as the functional angle and use the trigonometric functions of angle A to solve for the unknown elements of a right triangle.

The six trigonometric functions of angle A in Figure 11-3 or any other right triangle are:

$$\text{sine} = \frac{\text{opposite side}}{\text{hypotenuse}} \qquad \sin A = \frac{a}{c}$$

$$\text{cosine} = \frac{\text{adjacent side}}{\text{hypotenuse}} \qquad \cos A = \frac{b}{c}$$

$$\text{tangent} = \frac{\text{opposite side}}{\text{adjacent side}} \qquad \tan A = \frac{a}{b}$$

$$\text{cotangent} = \frac{\text{adjacent side}}{\text{opposite side}} \qquad \cot A = \frac{b}{a}$$

$$\text{secant} = \frac{\text{hypotenuse}}{\text{adjacent side}} \qquad \sec A = \frac{c}{b}$$

$$\text{cosecant} = \frac{\text{hypotenuse}}{\text{opposite side}} \qquad \text{cosec } A = \frac{c}{a}$$

These six functions give technicians ways to solve a right triangle where any side and one of the acute angles is known—if the technician knows how to determine the sine, cosine, tangent, cotangent, secant, or cosecant of an angle.

Fortunately, making this determination is a simple process in the age of computers. In the old days, technicians had to look up trigonometric functions in special tables such as the one in Figure 11-4. In Figure 11-4 to determine the sine of 20°30′, locate 20°30′ under the "degrees" column and move across to the right until you reach the corresponding number under the "sin" column. This number should be 0.3502. The same process is repeated to determine the cosine, tangent, cotangent, secant, and cosecant.

Degrees	Radians	sin	cos	tan	cot	sec	csc		
18° 00′	.3142	.3090	.9511	.3249	3.078	1.051	3.236	1.2566	**72° 00′**
10	171	118	502	281	047	052	207	537	50
20	200	145	492	314	018	053	179	508	40
30	.3229	.3173	.9483	.3346	2.989	1.054	3.152	1.2479	30
40	258	201	474	378	960	056	124	450	20
50	287	228	465	411	932	057	098	421	10
19° 00′	.3316	.3256	.9455	.3443	2.904	1.058	3.072	1.2392	**71° 00′**
10	345	283	446	476	877	059	046	363	50
20	374	311	436	508	850	060	021	334	40
30	.3403	.3338	.9426	.3541	2.824	1.061	2.996	1.2305	30
40	432	365	417	574	798	062	971	275	20
50	462	393	407	607	773	063	947	246	10
20° 00′	.3491	.3420	.9397	.3640	2.747	1.064	2.924	1.2217	**70° 00′**
10	520	449	387	673	723	065	901	188	50
20	549	475	377	706	699	066	878	159	40
30	.3578	.3502	.9367	.3739	2.675	1.068	2.855	1.2130	30
40	607	529	356	772	651	069	833	101	20
50	636	557	346	805	628	070	812	072	10
21°00′	.3665	.3584	.9336	.3839	2.605	1.071	2.790	1.2043	**69° 00′**
10	694	611	325	872	583	072	769	1.2014	50
20	723	638	315	906	560	074	749	985	40
30	.3752	.3665	.9304	.3939	2.539	1.075	2.729	1.1956	30
40	782	692	293	973	517	076	709	926	20
50	811	719	283	.4006	496	077	689	897	10
22° 00′	.3840	.3746	.9272	.4040	2.475	1.079	2.669	1.1868	**68° 00′**
10	869	773	261	074	455	080	650	839	50
20	898	800	250	108	434	081	632	810	40
30	.3927	.3827	.9239	.4142	2.414	1.082	2.613	1.1781	30
40	956	854	228	176	394	084	595	752	20
50	985	881	216	210	375	085	577	723	10
23° 00′	.4014	.3907	.9205	.4245	2.356	1.086	2.559	1.1694	**67° 00′**
10	043	934	194	279	337	088	542	665	50
20	072	961	182	314	318	089	525	636	40
30	.4102	.3987	.9171	.4348	2.300	1.090	2.508	1.1606	30
40	131	.4014	159	383	282	092	491	577	20
50	160	041	147	417	264	093	475	548	10
24° 00′	.4189	.4067	.9135	.4452	2.246	1.095	2.459	1.1519	**66° 00′**
10	218	094	124	487	229	096	443	490	50
20	247	120	112	522	211	097	427	461	40
30	.4276	.4147	.9100	.4557	2.194	1.099	2.411	1.1432	30
40	305	173	088	592	177	100	396	403	20
50	334	200	075	628	161	102	381	374	10
25° 00′	.4363	.4226	.9063	.4663	2.145	1.103	2.366	1.1345	**65° 00′**
10	392	253	051	699	128	105	352	316	50
20	422	279	038	734	112	106	337	286	40
30	.4451	.4305	.9026	.4770	2.097	1.108	2.323	1.1257	30
40	480	331	013	806	081	109	309	228	20
50	509	358	001	841	066	111	295	199	10
26° 00′	.4538	.4384	.8988	.4877	2.050	1.113	2.281	1.1170	**64° 00′**
10	567	410	975	913	035	114	268	141	50
20	596	436	962	950	020	116	254	112	40
30	.4625	.4462	.8949	.4986	2.006	1.117	2.241	1.1083	30
40	654	488	936	.5022	1.991	119	228	054	20
50	683	514	923	059	977	121	215	1.1025	10
27° 00′	.4712	.4540	.8910	.5095	1.963	1.122	2.203	1.0996	**63° 00′**
		cos	sin	cot	tan	csc	sec	Radians	Degrees

Figure 11-4 (Source: Sobel/Lerner, *Algebra and Trigonometry: A Pre-Calculus Approach*, 2/e, © 1985, p. 453. Reprinted by permission of Prentice-Hall, Inc., Englewood Cliffs, NJ.)

With modern electronic calculators, this process has been simplified considerably. The angle is entered in decimal form and the appropriate function button(s) is pressed. This done, the trigonometric function in question appears on the calculator's display.

PRACTICE PROBLEM SET 11-2

Use an electronic calculator (preferred) or the trigonometric table in Appendix A (Table A-9) to determine each function.

1. Sine 31°14′
2. Cosine 14°15′
3. Tangent 73°12′
4. Cotangent 9°08′
5. Secant 25°13′
6. Cosecant 24°09′
7. Sine 03°18′15″
8. Cosine 46°11′13″
9. Tangent 87°14′46″
10. Cotangent 15°15″

Solving Right Triangles Using the Sine Function

The sine function can be used when one acute angle and either the opposite side or the hypotenuse is known, as in Figure 11-5. In this right triangle, angle *A* and the hypotenuse are known. Angle *B* and side *a* are determined as follows:

Step 1

Solve for angle *B* by subtraction.

$$90 - 28 = 62°$$

Step 2

The sine function may be used to solve for the hypotenuse or the opposite side. Since the hypotenuse is known (15), the sine function is used. Solve for the opposite side (side *a*).

$$\sin A = \frac{a}{c}$$

$$\sin 28 = \frac{a}{15}$$

$$0.4695 = \frac{a}{15}$$

$$(15)(0.4695) = \frac{a}{15}(15)$$

$$7.04 = a$$

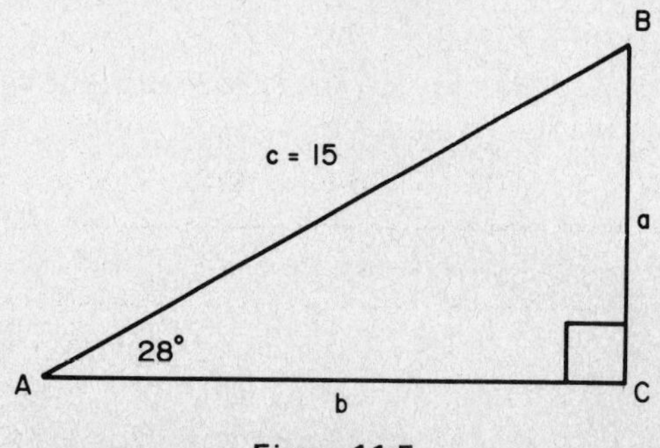

Figure 11-5

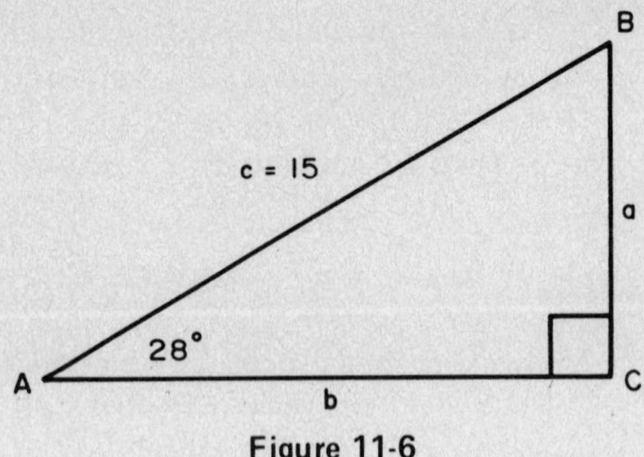

Figure 11-6

Solving Right Triangles Using the Cosine Function

The cosine function can be used when one acute angle and either the hypotenuse or the adjacent side are known, as in Figure 11-6. In this right triangle, angle A and the hypotenuse are known. Angle B and side b are determined as follows:

Step 1

Solve for angle B by subtraction.

$$90 - 28 = 62^\circ$$

Step 2

The cosine function may be used to solve for either the hypotenuse or the adjacent side. Since the hypotenuse is known (15), the cosine function is used to solve for the adjacent side (side b).

$$\cos A = \frac{b}{c}$$

$$\cos 28 = \frac{b}{c}$$

$$0.8829 = \frac{b}{15}$$

$$(15)(0.8829) = \frac{b}{15}(15)$$

$$13.24 = b$$

Solving Right Triangles Using the Tangent Function

The tangent function can be used when one acute angle and either the opposite or the adjacent side are known, as in Figure 11-7. In this right triangle, angle A and the adjacent side are known. Angle B and side a are determined as follows:

Step 1

Solve for angle B by subtraction.

$$90 - 22 = 68^\circ$$

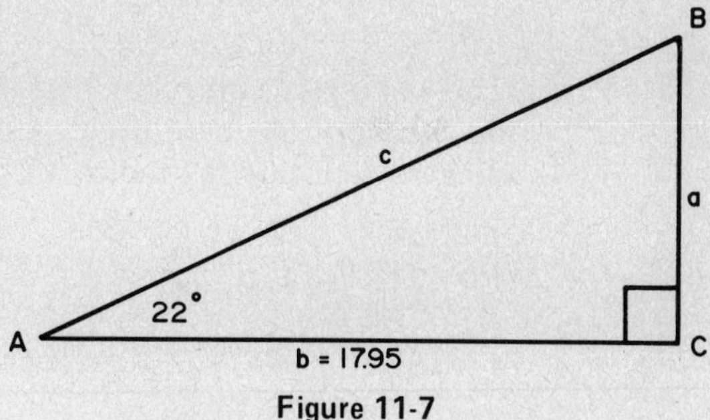

Figure 11-7

Step 2

The tangent function may be used to solve for either the opposite or the adjacent side. Since the adjacent side is known (17.95), the tangent function is used to solve for the opposite side (side a).

$$\tan A = \frac{a}{b}$$

$$\tan 22 = \frac{a}{17.95}$$

$$0.4040 = \frac{a}{17.95}$$

$$17.95(0.4040) = \frac{a}{17.95}(17.95)$$

$$7.25 = a$$

Solving Right Triangles Using the Cotangent Function

The cotangent function can be used when one acute angle and either the opposite or adjacent angle are known, as in Figure 11-8. In this right triangle, angle A and the opposite side are known. Angle B and side b are determined as follows:

Step 1

Solve for angle B by subtraction.

$$90 - 34 = 56^\circ$$

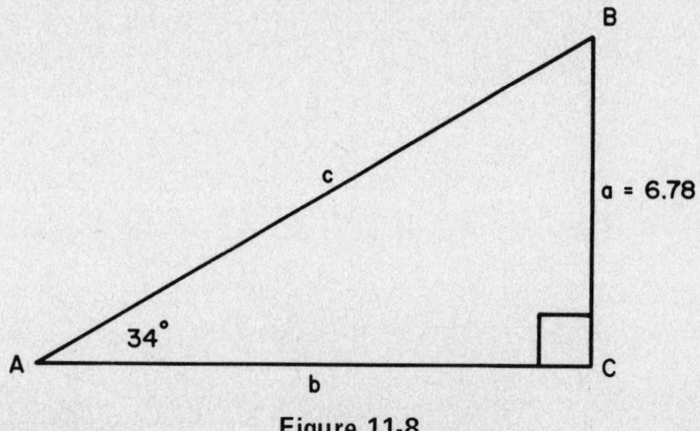

Figure 11-8

Step 2

The cotangent function may be used to solve for either the adjacent or the opposite side. Since the opposite side is known (6.78), the cotangent function is used to solve for the adjacent side (side b).

$$\cot A = \frac{b}{a}$$

$$\cot 34 = \frac{b}{6.78}$$

$$1.4826 = \frac{b}{6.78}$$

$$(6.78)(1.4826) = \frac{b}{6.78}(6.78)$$

$$10.05 = b$$

Solving Right Triangles Using the Secant Function

The secant function can be used when one acute angle and either the hypotenuse or the adjacent side are known, as in Figure 11-9. In this triangle, angle A and the adjacent angle are known. Angle B and the hypotenuse are determined as follows:

Step 1

Solve for angle B by subtraction.

$$90 - 23 = 67^\circ$$

Step 2

The secant function may be used to solve for either the hypotenuse or the adjacent side. Since the adjacent side is known (37.58), the secant function is used to solve for the hypotenuse (side c).

$$\sec A = \frac{c}{b}$$

$$\sec 23 = \frac{c}{37.58}$$

$$1.0864 = \frac{c}{37.58}$$

$$(37.58)(1.0864) = \frac{c}{37.58}(37.58)$$

$$40.83 = c$$

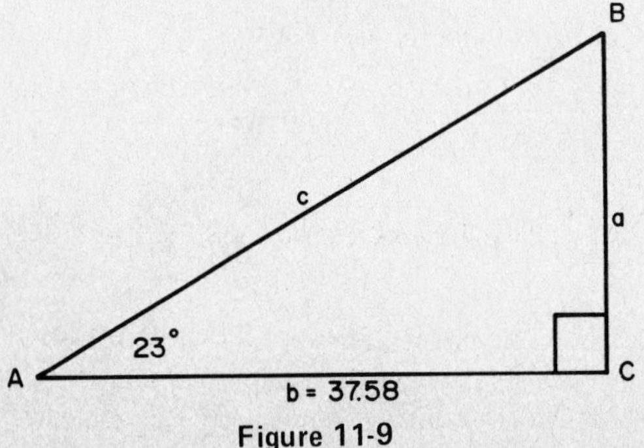

Figure 11-9

Solving Right Triangles Using the Cosecant Function

The cosecant function can be used when one acute angle and either the hypotenuse or the opposite side are known, as in Figure 11-10. In this right triangle, angle A and the opposite side are known. Angle B and the hypotenuse are determined as follows:

Step 1

Solve for angle B by subtraction.

$$90 - 37 = 53^\circ$$

Step 2

The cosecant function may be used to solve for either the hypotenuse or the opposite side. Since the opposite side is known (4.03), the cosecant function is used to solve for the hypotenuse (side c).

$$\text{cosec } A = \frac{c}{a}$$

$$\text{cosec } 37 = \frac{c}{4.03}$$

$$1.6616 = \frac{c}{4.03}$$

$$(4.03)(1.6616) = \frac{c}{4.03}(4.03)$$

$$6.70 = c$$

In all of the examples presented so far, one of the six trigonometric functions was used to solve for an unknown side, and subtraction was used for the unknown angle. A right triangle is not solved until all three sides are known.

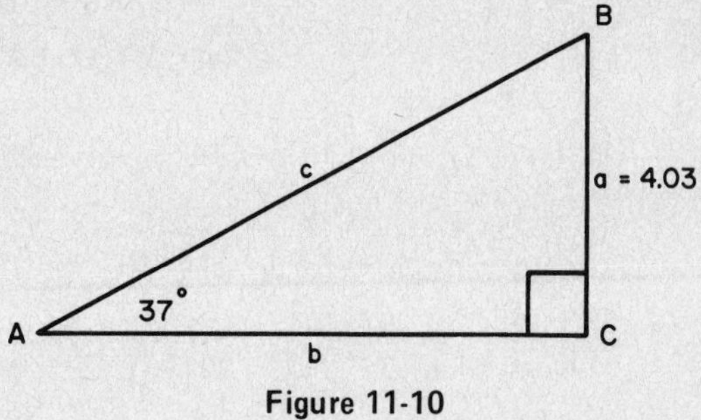

Figure 11-10

To solve for the still unknown side in each case presented so far, the known sides and another trigonometric function or the Pythagorean theorem can be used.

Using Trigonometric Functions to Solve for Angles

It is possible to use trigonometric functions to solve for unknown angles in a right triangle. This can be done when the two sides forming the angle are known, as in Figure 11-11.

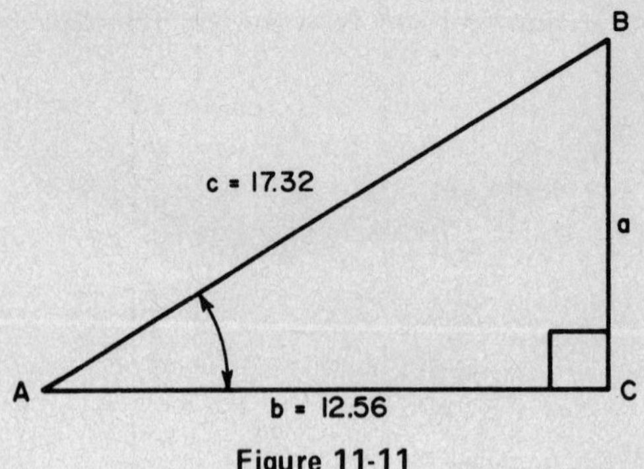

Figure 11-11

In this triangle, neither acute angle is known. However, the two sides that form angle A are known (side $b = 12.56$ and side $c = 17.32$). Side b is the adjacent side and side c is the hypotenuse. Consequently, any trigonometric function that includes the adjacent side and the hypotenuse can be used to solve for angle A. This means that the cosine and the secant functions can be used.

Using the cosine function, angle A can be determined as follows:

$$\cos A = \frac{b}{c}$$

$$\cos A = \frac{12.56}{17.32}$$

$$\cos A = 0.7252$$

$$\text{angle } A = 43.52^\circ \text{ or } 43^\circ 31'$$

The angle is determined by finding the angle whose cosine is 0.7252. This can be done using the "inverse" function on a calculator or by looking up 0.7252 in a trigonometric functions table such as Table A-9 in Appendix A.

PRACTICE PROBLEM SET 11-3

Refer to Figure 11-12 and use trigonometric functions to solve each right triangle.

1. $A = 12$, $a = 19.56$, $B = ?$, $b = ?$, $c = ?$
2. $A = ?$, $a = 7.51$, $B = 56$, $b = ?$, $c = ?$
3. $A = ?$, $a = ?$, $B = 67$, $b = ?$, $c = 71.20$
4. $A = 12$, $a = ?$, $B = ?$, $b = 19.51$, $c = ?$

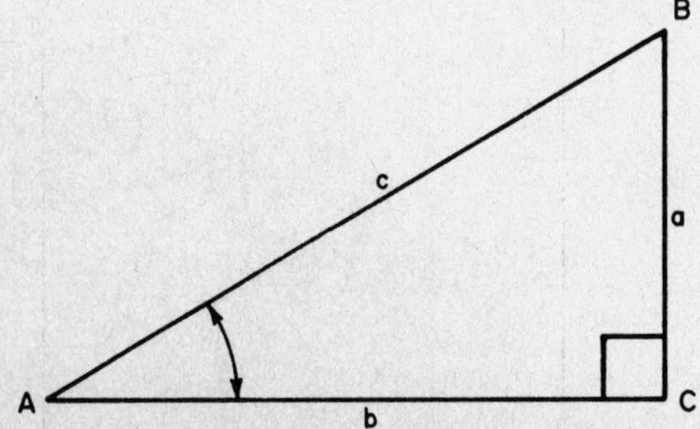

Figure 11-12

5. $A = ?$, $a = 3.48$, $B = 49$, $b = ?$, $c = ?$
6. $A = 11$, $a = ?$, $B = ?$, $b = ?$, $c = 128$
7. $A = 14$, $a = ?$, $B = ?$, $b = 15.03$, $c = ?$
8. $A = ?$, $a = ?$, $B = ?$, $b = 16.56$, $c = 19.75$
9. $A = ?$, $a = ?$, $B = ?$, $b = 17$, $c = 21$
10. $A = 18$, $a = 14$, $B = ?$, $b = ?$, $c = ?$

REVIEW PROBLEMS

Use the Pythagorean theorem to solve for the missing side in each right triangle.

1. $a = 7$, $b = 9$, $c = ?$
2. $a = 9$, $b = 11$, $c = ?$
3. $a = 10$, $b = 12$, $c = ?$
4. $a = 11$, $b = 13$, $c = ?$
5. $a = 14.5$, $b = 16.5$, $c = ?$
6. $a = ?$, $b = 10$, $c = 13$
7. $a = ?$, $b = 8$, $c = 16$
8. $a = 12.59$, $b = ?$, $c = 16.63$
9. $a = 5.04$, $b = ?$, $c = 8.12$
10. $a = ?$, $b = 8.14$, $c = 19.05$

Solve for the missing components in each right triangle.

11. $A = 13$, $a = 20.57$, $B = ?$, $b = ?$, $c = ?$
12. $A = ?$, $a = 8.58$, $B = 59$, $b = ?$, $c = ?$
13. $A = ?$, $a = ?$, $B = 64$, $b = ?$, $c = 72.48$
14. $A = 12.67$, $a = ?$, $B = 19.57$, $b = ?$, $c = ?$
15. $A = ?$, $a = 8.34$, $B = 59$, $b = ?$, $c = ?$
16. $A = 13.11$, $a = ?$, $B = ?$, $b = ?$, $c = 182$
17. $A = 16$, $a = ?$, $B = ?$, $b = 17.33$, $c = ?$
18. $A = ?$, $a = ?$, $B = ?$, $b = 18.57$, $c = 29.75$
19. $A = ?$, $a = ?$, $B = ?$, $b = 19$, $c = 27$
20. $A = 38$, $a = 24$, $B = ?$, $b = ?$, $c = ?$

12
TRIGONOMETRY: OBLIQUE TRIANGLES

An oblique triangle is any triangle that is not a right triangle. "Oblique" is a broad term that encompasses both acute and obtuse triangles. The Pythagorean theorem and trigonometric functions do not apply to oblique triangles. However, there are two commonly used methods for solving oblique triangles: the law of sines and the law of cosines.

SOLVING OBLIQUE TRIANGLES: LAW-OF-SINES METHOD

The *law of sines* states that "in any triangle, the sides are proportional to the sines of their opposite angles." This relationship can be translated into a formula that can be used for solving certain oblique triangles:

$$\frac{a}{\text{sine } A} = \frac{b}{\text{sine } B} = \frac{c}{\text{sine } C}$$

This formula can be used to solve oblique triangles in two cases:

1. When two angles and a side opposite one of them is known
2. When two sides and an angle opposite one of them are known

From the two cases above, it can be seen that *three* elements must be known in order to solve an oblique triangle using the law of sines. In selecting the two proportions from the law-of-sines formula that will be used, you must select a combination that includes the three elements that are known and the one element that is not.

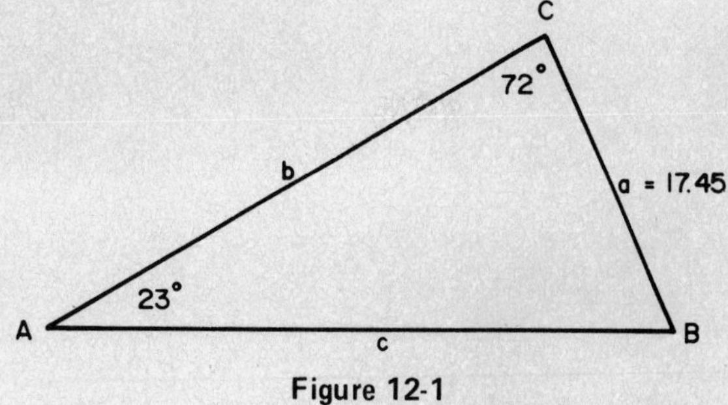

Figure 12-1

Solving Oblique Triangles: Given One Side and Two Angles

When two angles in an oblique triangle are known, the third angle can be calculated by subtracting the two known angles from 180°. The two unknown sides can be calculated using the law-of-sines formula as in the following example. Figure 12-1 contains an oblique triangle in which one side and two angles are known. Note that the angles may be given any letter designation (in capital letters), but the side opposite an angle must have the corresponding lowercase letter designation. To solve the oblique triangle in Figure 12-1, you must solve for angle B, side b, and side c.

Step 1

Solve for angle B by subtraction.

$$180 - 23 - 72 = 85°$$

Step 2

Solve for side b using the appropriate proportion combination from the law-of-sines formula.

$$\frac{a}{\text{sine } A} = \frac{b}{\text{sine } B}$$

$$\frac{17.45}{\text{sine } 23} = \frac{b}{\text{sine } 85}$$

$$\frac{17.45}{0.3907} = \frac{b}{0.9962}$$

$$0.3907b = 17.3837$$

$$b = 44.49$$

Step 3

Solve for side c using the appropriate proportion combination from the law-of-sines formula (remember to select combinations that use *given* rather than *calculated* information whenever possible).

$$\frac{a}{\text{sine } A} = \frac{c}{\text{sine } C}$$

$$\frac{17.45}{\text{sine } A} = \frac{c}{\text{sine } C}$$

$$\frac{17.45}{0.3907} = \frac{c}{0.9511}$$

$$0.3907c = 16.60$$
$$c = 42.48$$

Solving Oblique Triangles: Given Two Sides and an Angle

When two sides and an angle opposite one of them are known (see Figure 12-2), the law-of-sines formula can be used to solve the triangle as follows:

1. Solve for the angle opposite the other known side using the appropriate combination from the law-of-sines formula.
2. Solve for the third angle by subtraction.
3. Solve for the missing side using the appropriate combination from the law-of-sines formula.

This process is illustrated in the following example: In order to solve the oblique triangle in Figure 12-2, you must solve for angle A, angle C, and side c.

Step 1

Solve for angle A using the appropriate combination from the law-of-sines formula.

$$\frac{a}{\text{sine } A} = \frac{b}{\text{sine } B}$$
$$\frac{15}{\text{sine } A} = \frac{18}{\text{sine } 22}$$
$$\frac{15}{\text{sine } A} = \frac{18}{0.3746}$$
$$(18)(\text{sine } A) = 5.619$$
$$(18)\frac{\text{sine } A}{18} = \frac{5.619}{18}$$
$$\text{sine } A = 0.3122$$
$$A = 18.1899^\circ$$
$$A = 18^\circ 11' 23''$$

Step 2

Solve for angle C by subtraction.

$$C = 180 - 22 - 18.1899$$
$$C = 139.8101$$
$$C = 139^\circ 48' 37''$$

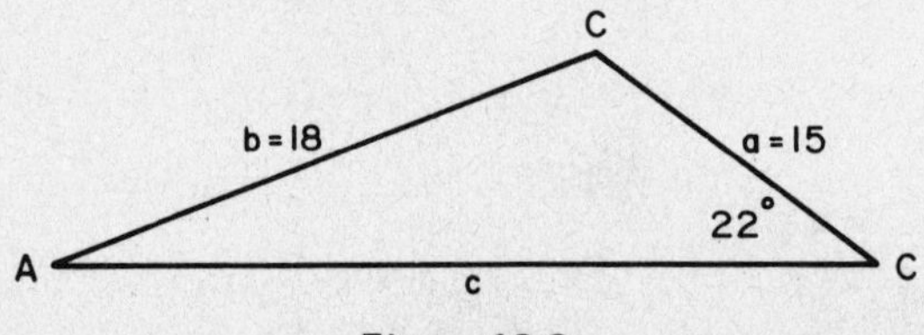

Figure 12-2

Step 3

Solve for side c using the appropriate combination from the law-of-sines formula.

$$\frac{a}{\text{sine } A} = \frac{c}{\text{sine } C}$$

$$\frac{15}{\text{sine } 18.1899} = \frac{c}{\text{sine } 139.8103}$$

$$\frac{15}{0.3122} = \frac{c}{0.6453}$$

$$0.3222c = 9.6795$$

$$\frac{0.3122c}{0.3122} = \frac{9.6795}{0.3122}$$

$$c = 31$$

PRACTICE PROBLEM SET 12-1

Use the law of sines to solve each oblique triangle. Sketch a triangle from the information given before beginning the solution.

1. $A = ?$, $B = 64$, $C = 50$, $a = 32$, $b = ?$, $c = ?$
2. $A = 41$, $B = 82$, $C = ?$, $a = ?$, $b = 31$, $c = ?$
3. $A = 72$, $B = ?$, $C = 57$, $a = ?$, $b = 10.12$, $c = ?$
4. $A = ?$, $B = 7$, $C = 110$, $a = ?$, $b = ?$, $c = 12.51$
5. $A = 83$, $B = 22$, $C = ?$, $a = 6$, $b = ?$, $c = ?$
6. $A = 69$, $B = ?$, $C = ?$, $a = 39$, $b = 29$, $c = ?$
7. $A = 33$, $B = ?$, $C = ?$, $a = 7$, $b = 11$, $c = ?$
8. $A = 75$, $B = ?$, $C = ?$, $a = 24$, $b = 17$, $c = ?$
9. $A = ?$, $B = ?$, $C = 65$, $a = ?$, $b = 19$, $c = 22$
10. $A = 31$, $B = ?$, $C = ?$, $a = 56$, $b = 14$, $c = ?$

SOLVING OBLIQUE TRIANGLES: LAW-OF-COSINES METHOD

The *law of cosines* states that "the square of any side of a triangle is equal to the sum of the squares of the two other sides minus twice the product of these two sides multiplied by the cosine of their included angle." This statement can be translated into a formula with three derivations. These derivations of the law-of-cosines formula can be used for solving certain oblique triangles:

$$a^2 = b^2 + c^2 - 2bc \text{ cosine } A$$

$$b^2 = a^2 + c^2 - 2ac \text{ cosine } B$$

$$c^2 = a^2 + b^2 - 2ab \text{ cosine } C$$

These formulas can be used to solve oblique triangles in two cases:

1. When two sides and the included angle are known
2. When three sides are known

Solving Oblique Triangles: Given Two Sides and the Included Angle

When two sides and their included angle are known, the law of cosines can be used to find the unknown side. Then the law of sines can be used to find the second angle. Subtraction can be used to find the third angle.

Figure 12-3 contains an oblique triangle in which two sides and the included angle are known. Angle $B = 89°$, side $a = 112$, and side $c = 137$. The unknown components would be determined as follows:

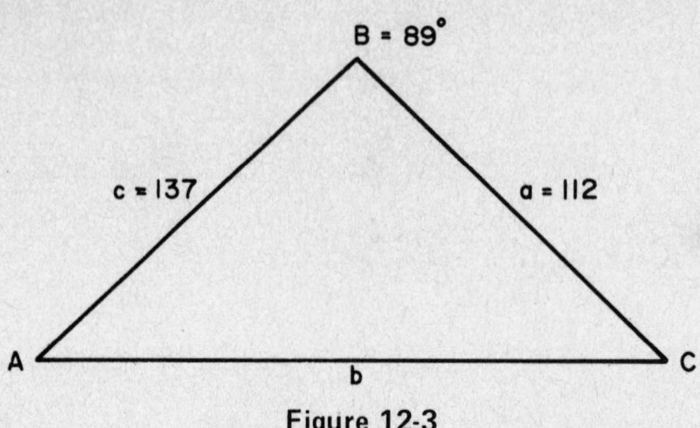

Figure 12-3

Step 1

Use the law-of-cosines formula to solve for side b.

$$b^2 = a^2 + c^2 - 2ac \text{ cosine } B$$
$$b^2 = 112^2 + 137^2 - 2(112)(137)(0.0174524)$$
$$b^2 = 12{,}544 + 18{,}769 - (30{,}688)(0.0174524)$$
$$b^2 = 31{,}313 - 535.57925$$
$$b^2 = 30{,}777.421$$
$$b^2 = 175.43$$

Step 2

Use the law-of-sines formula to solve for either angle A or angle C.

$$\frac{a}{\text{sine } A} = \frac{b}{\text{sine } B}$$
$$\frac{112}{\text{sine } A} = \frac{175.43}{0.9998477}$$
$$\text{sine } A\,(175.43) = 111.98$$
$$\text{sine } A\,\frac{175.43}{175.43} = \frac{111.98}{175.43}$$
$$\text{sine } A = 0.6383341$$
$$A = 39.667705$$
$$A = 39°40'04''$$

Step 3

Solve for the missing angle (C) by subtraction.

$$C = 180 - 89 - 39°40'04''$$
$$C = 51°19'56''$$

Solving Oblique Triangles: Given Three Sides

When three sides of an oblique triangle are known, the law of cosines can be used to find the unknown angles. To make the calculations simpler, the three derivatives of the law-of-cosines formula can be rearranged as follows:

$$\text{cosine } A = \frac{b^2 + c^2 - a^2}{2bc}$$

$$\text{cosine } B = \frac{a^2 + c^2 - b^2}{2ac}$$

$$\text{cosine } C = \frac{a^2 + b^2 - c^2}{2ab}$$

Figure 12-4 contains an oblique triangle in which three sides are known. The triangle would be solved as follows:

Step 1

Solve for angle A using the appropriate derivation of the law-of-cosines formula.

$$\text{cosine } A = \frac{b^2 + c^2 - a^2}{2bc}$$

$$\text{cosine } A = \frac{14^2 + 12^2 - 23^2}{2(14)(12)}$$

$$\text{cosine } A = \frac{340 - 529}{336}$$

$$\text{cosine } A = \frac{-189}{336}$$

$$\text{cosine } A = -0.5625$$

$$A = 124.2289$$

$$A = 124°13'44''$$

Step 2

Solve for angle B using the appropriate derivation of the law-of-cosines formula.

$$\text{cosine } B = \frac{a^2 + c^2 - b^2}{2ac}$$

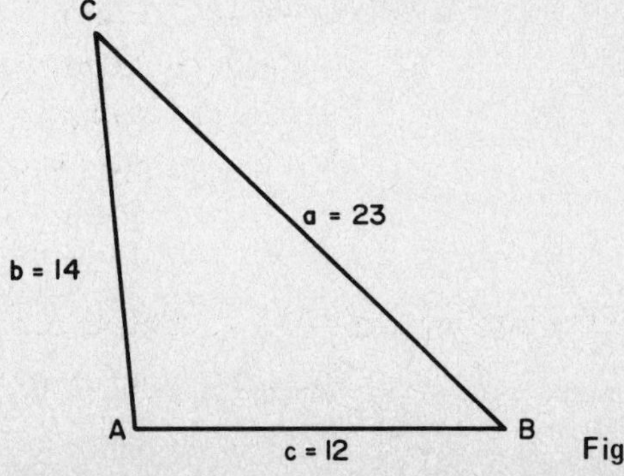

Figure 12-4

$$\text{cosine } B = \frac{23^2 + 12^2 - 14^2}{2(23)(12)}$$

$$\text{cosine } B = \frac{529 + 144 - 196}{552}$$

$$\text{cosine } B = \frac{447}{552}$$

$$\text{cosine } B = 0.8641304$$

$$B = 30.21644$$

$$B = 30°12'59''$$

Step 3

Solve for angle C using the appropriate derivation of the law-of-cosines formula.

$$\text{cosine } C = \frac{a^2 + b^2 - c^2}{2ab}$$

$$\text{cosine } C = \frac{23^2 + 14^2 - 12^2}{2(23)(14)}$$

$$\text{cosine } C = \frac{529 + 196 - 144}{644}$$

$$\text{cosine } C = \frac{581}{644}$$

$$\text{cosine } C = 0.9021739$$

$$C = 25.554694$$

$$C = 25°33'17''$$

After using the law of cosines to solve for all three angles, it is a good idea to make sure that your answers equal 180°. To check, simply add them together:

$$\begin{array}{r} 124°\ 13'\ 44'' \\ 30°\ 12'\ 59'' \\ 25°\ 13'\ 59'' \\ \hline 180°\ 00'\ 00'' \end{array}$$

PRACTICE PROBLEM SET 12-2

Use the law of cosines to solve each oblique triangle. Sketch a triangle from the given information before beginning the solution.

1. $A = ?$, $B = ?$, $C = 38°$, $a = 15$, $b = 11$, $c = ?$
2. $A = ?$, $B = ?$, $C = 118°$, $a = 26$, $b = 14$, $c = ?$
3. $A = 75°$, $B = ?$, $C = ?$, $a = ?$, $b = 20$, $c = 16$
4. $A = 61°$, $B = ?$, $C = ?$, $a = ?$, $b = 60$, $c = 48$
5. $A = 43°$, $B = ?$, $C = ?$, $a = ?$, $b = 45$, $c = 36$
6. $A = ?$, $B = ?$, $C = ?$, $a = 42$, $b = 32$, $c = 22$
7. $A = ?$, $B = ?$, $C = ?$, $a = 23$, $b = 26$, $c = 35$
8. $A = ?$, $B = ?$, $C = ?$, $a = 55$, $b = 72$, $c = 88$
9. $A = ?$, $B = ?$, $C = ?$, $a = 90$, $b = 98$, $c = 148$
10. $A = ?$, $B = ?$, $C = ?$, $a = 81$, $b = 142$, $c = 173$

REVIEW PROBLEMS

Use the law of sines to solve each oblique triangle.

1. $A = ?$, $B = 66$, $C = 52$, $a = 34$, $b = ?$, $c = ?$

2. $A = 44$, $B = 85$, $C = ?$, $a = ?$, $b = 34$, $c = ?$

3. $A = 76$, $B = ?$, $C = 61$, $a = ?$, $b = 14.16$, $c = ?$

4. $A = ?$, $B = 12$, $C = 115$, $a = ?$, $b = ?$, $c = 17.56$

5. $A = 89$, $B = 13$, $C = ?$, $a = 12$, $b = ?$, $c = ?$

6. $A = 79$, $B = ?$, $C = ?$, $a = 49$, $b = 59$, $c = ?$

7. $A = 44$, $B = ?$, $C = ?$, $a = 18$, $b = 22$, $c = 18$

8. $A = 72$, $B = ?$, $C = ?$, $a = 21$, $b = 14$, $c = ?$

9. $A = ?$, $B = ?$, $C = 65$, $a = ?$, $b = 19$, $c = 22$

10. $A = 13$, $B = ?$, $C = ?$, $a = 65$, $b = 41$, $c = ?$

Use the law of cosines to solve each oblique triangle.

11. $A = ?$, $B = ?$, $C = 40$, $a = 17$, $b = 13$, $c = ?$

12. $A = ?$, $B = ?$, $C = 120$, $a = 29$, $b = 17$, $c = ?$

13. $A = 79$, $B = ?$, $C = ?$, $a = ?$, $b = 24$, $c = 20$

14. $A = 66$, $B = ?$, $C = ?$, $a = ?$, $b = 65$, $c = 53$

15. $A = 49$, $B = ?$, $C = ?$, $a = ?$, $b = 51$, $c = 42$

16. $A = ?$, $B = ?$, $C = ?$, $a = 24$, $b = 23$, $c = 22$

17. $A = ?$, $B = ?$, $C = ?$, $a = 33$, $b = 36$, $c = 45$

18. $A = ?$, $B = ?$, $C = ?$, $a = 51$, $b = 69$, $c = 85$

19. $A = ?$, $B = ?$, $C = ?$, $a = 88$, $b = 96$, $c = 146$

20. $A = ?$, $B = ?$, $C = ?$, $a = 71$, $b = 132$, $c = 163$

13

TRIGONOMETRY: AREA OF TRIANGLES

Technicians are frequently required to calculate the area of triangles or the area of other shapes that involve triangles. There are several ways of calculating the area of a triangle. One method applies solely to right triangles and another method applies to all triangles but is used mostly for oblique triangles. Technicians need to be familiar with both of these methods.

CALCULATING THE AREA OF RIGHT TRIANGLES

A right triangle is actually half of a rectangle. Consequently, the formula used for calculating the area of a rectangle can be altered slightly and used for calculating the area of a right triangle. The formula for calculating the area of a rectangle is:

$$\text{area of a rectangle} = \text{base} \times \text{height} \qquad \text{or} \qquad A = bh$$

Since a right triangle is a rectangle halved diagonally, the formula above can be divided in half and used for calculating the area of a right triangle.

$$A = \frac{1}{2}bh \qquad \text{or} \qquad A = \frac{1}{2}ab$$

Figure 13-1 is a right triangle in which two sides (side a and side b) are known. The area of this right triangle would be calculated as follows:

Step 1

Set up the area formula.

$$A = \frac{1}{2}ab$$

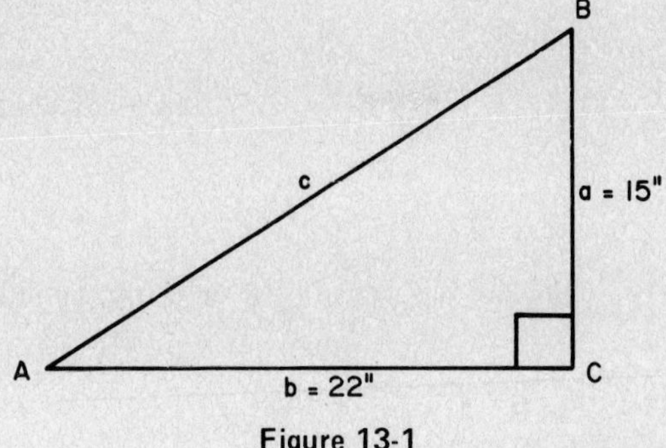

Figure 13-1

Step 2

Substitute numbers for literals and solve.

$$A = \frac{1}{2}(15)(22)$$

$$A = 165 \text{ square inches}$$

Figure 13-2 is a right triangle in which one side (side A) and the hypotenuse are known. A situation such as this requires an intermediate procedure in which the Pythagorean theorem is used to calculate side b since the hypotenuse is not used in the area formula. The area of this right triangle would be calculated as follows:

Step 1

Solve for side b using the Pythagorean theorem.

$$A^2 + b^2 = C^2$$

$$14^2 + b^2 = 36^2$$

$$196 + b^2 = 1{,}296$$

$$196 - 196 + b^2 = 1{,}296 - 196$$

$$b^2 = 1{,}100$$

$$b = 33.17$$

Step 2

Set up the area formula, substitute numbers for literals, and solve.

$$A = \frac{1}{2}ab$$

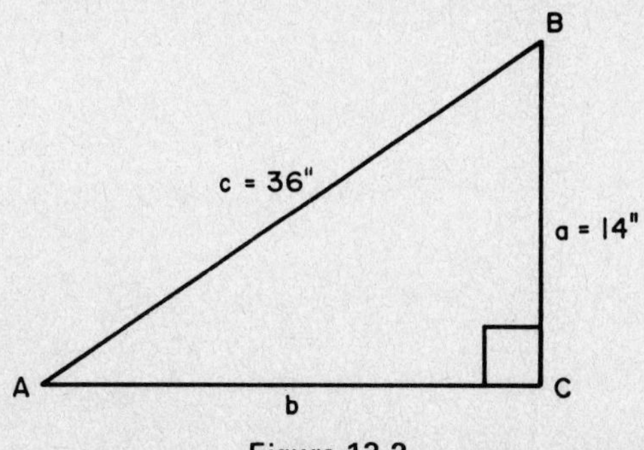

Figure 13-2

$$A = \frac{1}{2}(14)(33.17)$$

$$A = 239.19 \text{ square inches}$$

PRACTICE PROBLEM SET 13-1

Calculate the area of each right triangle (all dimensions are inches).

1. $a = 12, \quad b = 15$
2. $a = 14, \quad b = 17$
3. $a = 23, \quad b = 27$
4. $a = 7.95, \quad b = 12.04$
5. $a = 6.01, \quad b = 9.46$
6. $a = 10, \quad c = 22$
7. $a = 15.51, \quad c = 46$
8. $a = 20.75, \quad c = 39.33$
9. $b = 13.44, \quad c = 17.87$
10. $b = 19.64, \quad c = 24.43$

CALCULATING THE AREA OF OBLIQUE TRIANGLES

Calculating the area of oblique triangles is not as simple as calculating the area of right triangles, but it is not a difficult process. The formula used for calculating the areas of oblique triangles has three derivations.

$$A = \frac{1}{2}ab \text{ sine } C$$

$$A = \frac{1}{2}ac \text{ sine } B$$

$$A = \frac{1}{2}bc \text{ sine } A$$

To use one of the three derivations of this formula, two sides and their included angle must be known. When the needed information is not known, it must be calculated before the formula can be applied.

Figure 13-3 is an oblique triangle in which two sides (sides b and c) and their included angle (angle A) are known. The area of this oblique triangle would be calculated as follows:

Step 1

Select the appropriate derivation of the formula based on the information known.

$$A = \frac{1}{2}bc \text{ sine } A$$

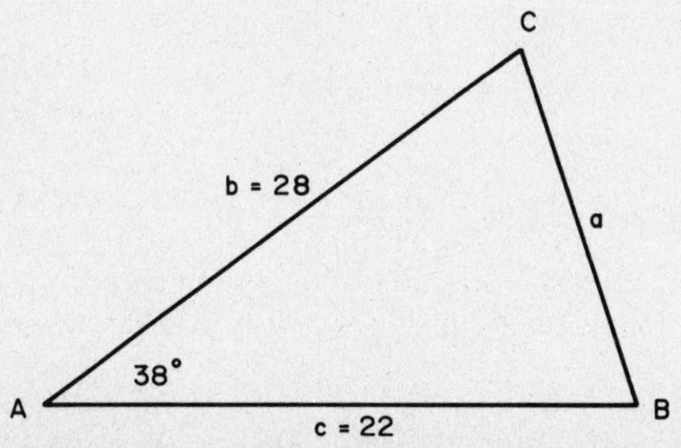

Figure 13-3

Step 2

Substitute numbers for literals and solve.

$$A = \frac{1}{2}(28)(22)(0.6156615)$$

$$A = 189.62 \text{ square inches}$$

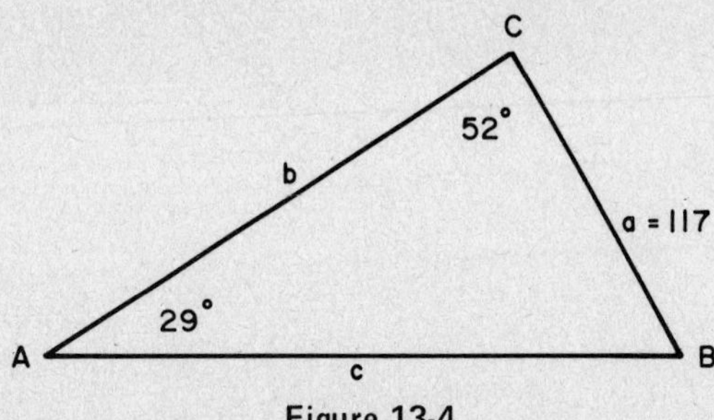

Figure 13-4

Figure 13-4 is an oblique triangle in which two angles and a side are known. Before the area formula can be applied, a second side must be solved. The area of this oblique triangle would be calculated as follows:

Step 1

Solve for side c using the law of sines.

$$\frac{a}{\text{sine } a} = \frac{c}{\text{sine } c}$$

$$\frac{117}{0.48480} = \frac{c}{0.7880}$$

$$190.18 = c$$

Step 2

Calculate angle B by subtraction.

$$\text{angle } B = 180 - 29 - 52$$

$$\text{angle } B = 99°$$

Step 3

Use the appropriate derivation of the area formula to calculate the area.

$$A = \frac{1}{2}ac \text{ sine } B$$

$$A = \frac{1}{2}(117)(190.18)(0.9876883)$$

$$A = 10{,}988.56 \text{ square inches}$$

Figure 13-5 is an oblique triangle in which two sides and an angle opposite one of them are known. Before the area formula can be applied, the angle included by the known side must be calculated. The area of oblique triangle in Figure 13-5 would be calculated as follows:

Step 1

Use the law of sines to calculate angle C.

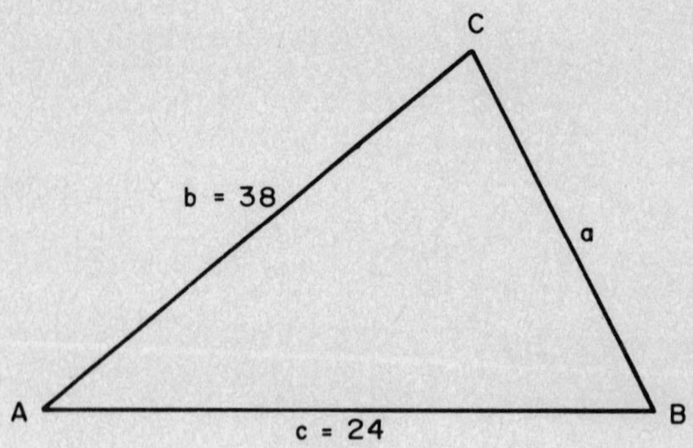

Figure 13-5

$$\frac{b}{\text{sine } B} = \frac{c}{\text{sine } C}$$

$$\frac{38}{0.8480} = \frac{24}{\text{sine } C}$$

$$0.5356093 = \text{sine } C$$

$$32.39 = C$$

$$32°23'07'' = C$$

Step 2

Solve for angle A (the included angle of the known sides) by subtraction.

$$\text{angle } A = 180 - 58 - (32°23'07'')$$

$$\text{angle } A = 89°36'53''$$

Step 3

Use the appropriate derivation of the area formula to calculate the area.

$$A = \frac{1}{2}bc \text{ sine } A$$

$$A = \frac{1}{2}(38)(24)(0.9999774)$$

$$A = 455.99 \text{ square inches}$$

PRACTICE PROBLEM SET 13-2

Calculate the area of each oblique triangle. Set up the problems as in the examples in Figures 13-3 through 13-5.

1. $A = 14°$, $b = 10$, $c = 12$
2. $A = 23°$, $b = 14$, $c = 17$
3. $A = 32°$, $b = 15.95$, $c = 18.01$
4. $B = 58°$, $a = 9.72$, $c = 14$
5. $B = 60°$, $a = 6.65$, $c = 13.91$
6. $A = 18°$, $C = 42°$, $a = 15$
7. $A = 24°$, $C = 46°$, $a = 76$
8. $A = 31°$, $B = 66°$, $b = 110$
9. $B = 64°$, $b = 55$, $c = 45$
10. $B = 43°$, $b = 91.14$, $c = 78.25$

REVIEW PROBLEMS

Calculate the area of each right triangle.

1. $a = 13$, $b = 16$
2. $a = 15$, $c = 18$
3. $a = 24$, $b = 29$
4. $a = 8.96$, $c = 12.55$

5. $a = 7.01$, $b = 19.46$
6. $a = 11$, $c = 23$
7. $a = 16.21$, $b = 36$
8. $a = 21.85$, $c = 49.34$
9. $a = 15.43$, $b = 17.78$
10. $a = 29.64$, $c = 34.34$

Calculate the area of each oblique triangle.

11. $A = 15$, $b = 11$, $c = 13$
12. $A = 25$, $b = 16$, $c = 19$
13. $A = 36$, $b = 16.86$, $c = 20.01$
14. $A = 16$, $C = 43$, $a = 13$
15. $A = 44$, $C = 46$, $a = 75$
16. $A = 21$, $B = 56$, $b = 110$
17. $A = 34$, $C = 45$, $a = 76$
18. $A = 51$, $B = 86$, $b = 91$
19. $B = 56$, $a = 19.76$, $c = 24$
20. $B = 74$, $b = 65$, $c = 55$

PART 3
Applying Math in the Machine Trades

14
APPLYING ARITHMETIC IN THE MACHINE TRADES

INTRODUCTION TO PART 3

Part 3 is devoted entirely to applying the knowledge and skills developed in Parts 1 and 2. Each chapter consists of real-world problems that require one or more mathematical procedures. The purpose of Part 3 is to give students the opportunity to apply their math skills as they will on the job.

In attempting to solve the problems in Part 3, students are encouraged to refer to Appendix A as frequently as they deem necessary. This will further reinforce their understanding of the world of work since practicing technicians must continually refer to a variety of references in completing their daily tasks on the job.

This is an application chapter. It consists solely of on-the-job-oriented problems which require you to apply what you have learned in earlier chapters. The problems in this chapter apply to what was learned in Chapters 1 to 5. You are encouraged to use Chapters 1 to 5 and all appended material for reference purposes in solving the problems in this chapter.

APPLICATION PROBLEM 1

You are required to cut several lengths off a piece of .75-in bar stock that is 72 in. long. If you cut off lengths of 13, 5, 7, and 14 in., how much of the original bar stock will be left?

APPLICATION PROBLEM 2

Refer to Figure 14-1 in calculating the following dimensions.

$$A = ?$$
$$B = ?$$

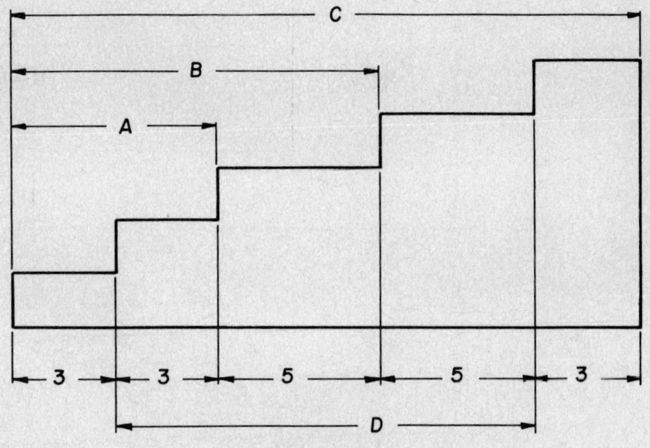

Figure 14-1

$C = ?$

$D = ?$

APPLICATION PROBLEM 3

Refer to Figure 14-2 in calculating the following dimensions.

$A = ?$

$B = ?$

$C = ?$

$D = ?$

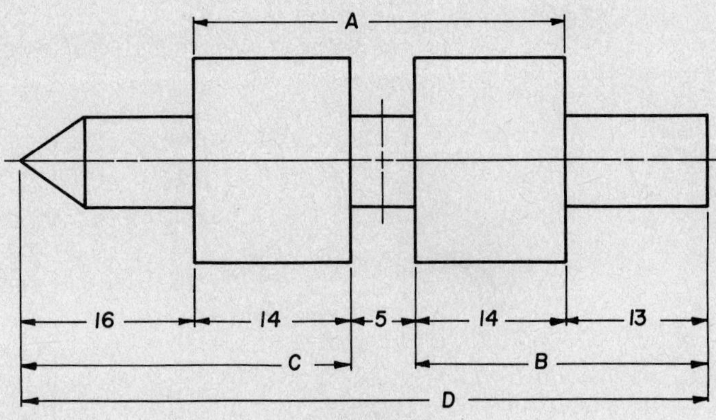

Figure 14-2

APPLICATION PROBLEM 4

Refer to Figure 14-3 in calculating the following distances.

(a) The distance from hole A to hole C.
(b) The distance from hole A to hole D.
(c) The overall length of the piece.

APPLICATION PROBLEM 5

You are taking inventory in the supply room and make the following counts:

15 Hex bolts

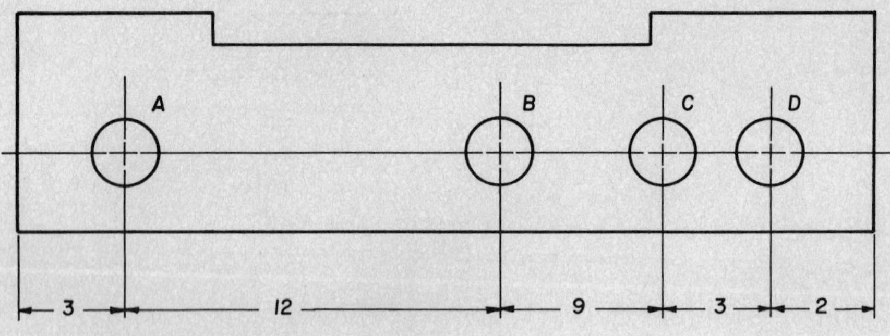

Figure 14-3

12 Square bolts
13 Heavy hex bolts
9 Capscrews
74 Heavy hex screws
16 Hex nuts
17 Hex jam nuts
24 Heavy hex nuts
16 Heavy hex jam nuts

(a) How many bolts do you have?
(b) How many screws do you have?
(c) How many hex nuts do you have?
(d) How many total items do you have?

APPLICATION PROBLEM 6

Refer to Figure 14-4 to calculate the diameter of hole A through the workpiece.

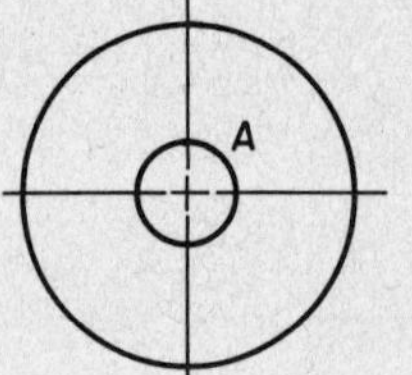

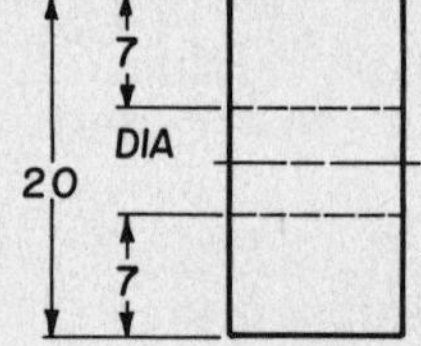

Figure 14-4

APPLICATION PROBLEM 7

Refer to Figure 14-5 to calculate dimensions X and Y.

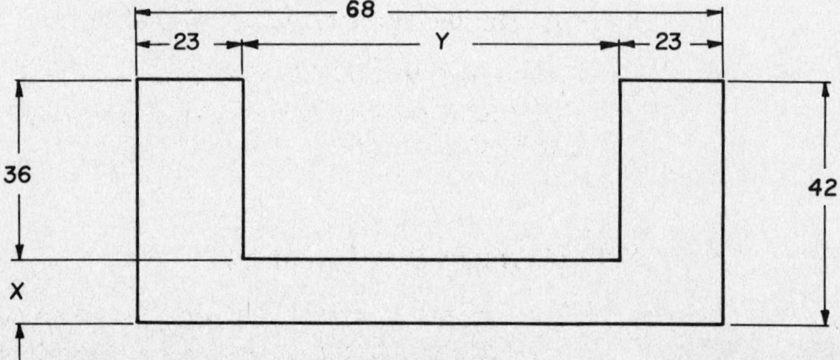

Figure 14-5

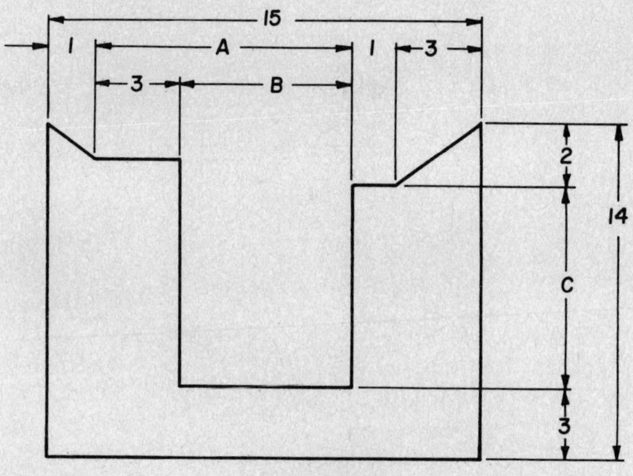

Figure 14-6

APPLICATION PROBLEM 8

Refer to Figure 14-6 to calculate dimensions *A*, *B*, and *C*.

APPLICATION PROBLEM 9

You have 108 pieces of bar stock to thread. Before lunch you threaded 9 pieces. So far after lunch you have threaded 7 more. How many do you have left to do?

APPLICATION PROBLEM 10

Total the labor hours that were required to complete Job 1715.

Job Number 1715

Operation	*Time*
Setup	15 hours
Turning	119 hours
Milling	217 hours
Drilling	47 hours
Grinding	16 hours
Heat treating	12 hours
Painting	8 hours

APPLICATION PROBLEM 11

You need to cut 15 pieces of bar stock, each 11 in. long. What is the total length in inches of the stock you will need to take from the storage bin?

APPLICATION PROBLEM 12

The original plate in Figure 14-7 weighed 80 ounces before the holes were drilled. Examine Figure 14-7 and calculate the weight of the plate after the holes are drilled.

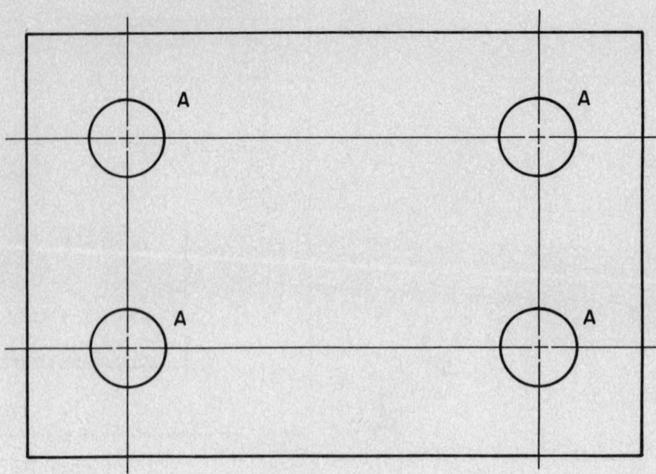

HOLE "A" WEIGHS 9 OUNCES

Figure 14-7

APPLICATION PROBLEM 13

You have 16 workpieces to machine. If each piece requires the hours set forth below, what is the total amount of time needed to complete all 16?

Operation	*Time*
Turning	3 hours
Milling	7 hours
Drilling	2 hours

APPLICATION PROBLEM 14

Your company has ordered 15 new direct numerical control (DNC) machining centers. The cost of shipping them depends on their weight. The cost is $2.00 per pound. How much will it cost to ship all of them if each weighs 3,372 pounds?

APPLICATION PROBLEM 15

A manufacturing plant produces diesel engines for trucks and buses. It produces two engines per hour, working one 8-hour shift per day. How many engines can it produce in a 5-day week if production time is increased to three 8-hour shifts per day?

APPLICATION PROBLEM 16

Refer to Figure 14-8 to calculate the dimension *X*. Each *X* is the same length.

APPLICATION PROBLEM 17

How many pieces 6 in. long can be cut from the bar stock in Figure 14-9?

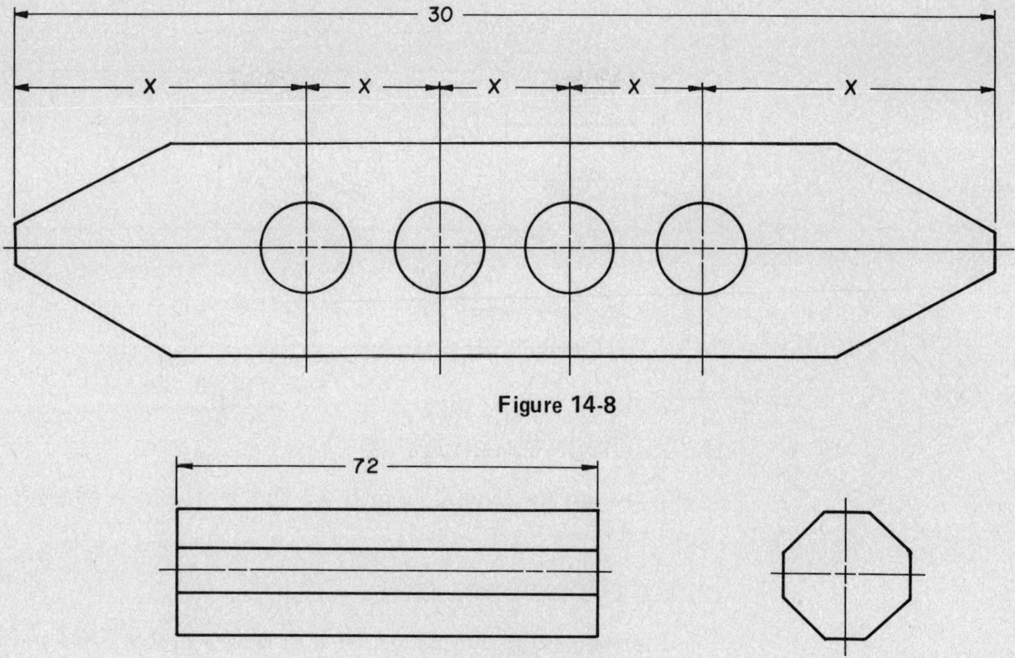

Figure 14-8

Figure 14-9

APPLICATION PROBLEM 18

Your company has dispatched you to accompany a truck driver and collect the money from selling a truckload of scrap steel. You are to collect $6.00 per ton for the scrap, and your truck contains 10 tons of scrap. (2,000 pounds = 1 ton.) You are to leave one-fifth of the scrap at your first stop and the rest at the second stop.

(a) How many pounds of scrap should you leave at the first stop, and how much money should you collect?

(b) How many pounds of scrap should you leave at the second stop, and how much should you collect?

APPLICATION PROBLEM 19

A multiple-spindle drill press can drill 72 holes per minute. How many holes can it drill in an 8-hour shift?

APPLICATION PROBLEM 20

A box of washers weighs 150 pounds. Each washer weighs 3 ounces. If the box weighs 5 pounds when empty, how many washers does it contain? (1 pound = 16 ounces.)

APPLICATION PROBLEM 21

Refer to Figure 14-10 and solve for dimension *X*.

APPLICATION PROBLEM 22

Refer to Figure 14-10 and solve for dimension *Y*.

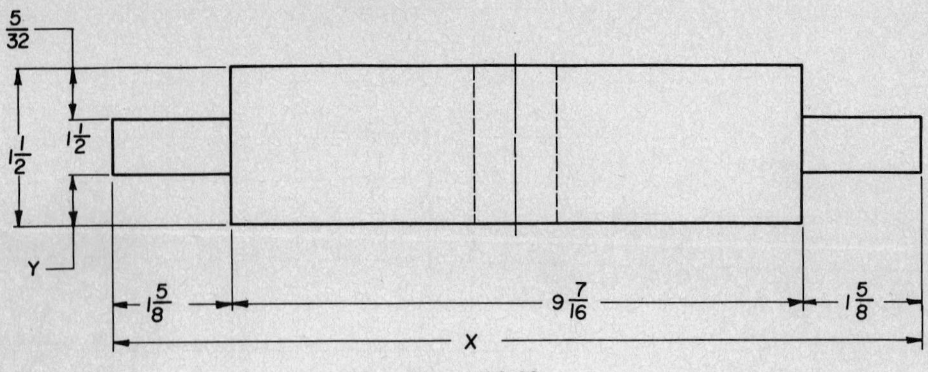

Figure 14-10

APPLICATION PROBLEM 23

What is the overall length of the workpiece (dimension D) in Figure 14-11?

APPLICATION PROBLEM 24

Increase each dimension in Figure 14-11 by 7/64 in. What are the new dimensions for the workpiece?

$A = ?$
$B = ?$
$C = ?$
$D = ?$

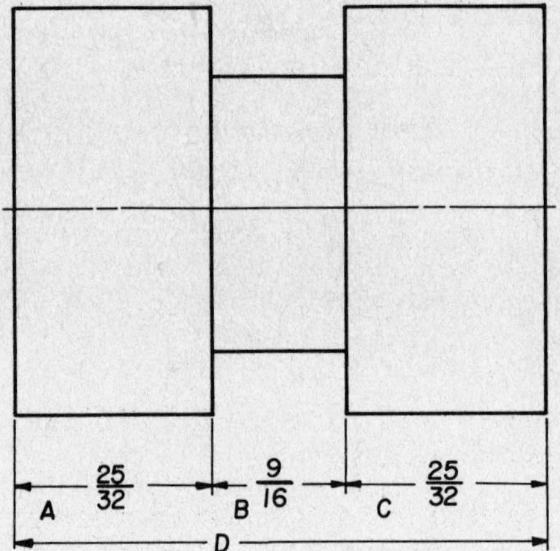

Figure 14-11

APPLICATION PROBLEM 25

You are required to machine a total of 47 slotted plates in five different configurations. Figure 14-12 contains a drawing and chart to guide you. Using the information given, calculate the overall dimension (B) for each configuration.

plate 1 = ?
plate 2 = ?
plate 3 = ?
plate 4 = ?
plate 5 = ?

PLATE NUMBER	CENTERLINE DISTANCE A BETWEEN HOLES	NUMBER REQUIRED
1	$2\frac{13}{16}$	15
2	$1\frac{5}{8}$	10
3	$3\frac{3}{16}$	5
4	$2\frac{25}{64}$	10
5	$1\frac{5}{32}$	7

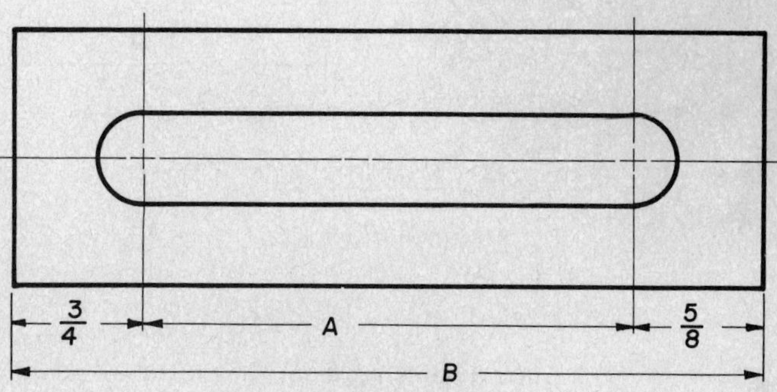

Figure 14-12

APPLICATION PROBLEM 26

Recalculate the overall dimensions for the plates in Figure 14-12 if 5/64 in. must be subtracted from each.

plate 1 = ?
plate 2 = ?
plate 3 = ?
plate 4 = ?
plate 5 = ?

APPLICATION PROBLEM 27

Calculate dimension X in Figure 14-13.

APPLICATION PROBLEM 28

What is the length of the bolt under the head in Figure 14-13?

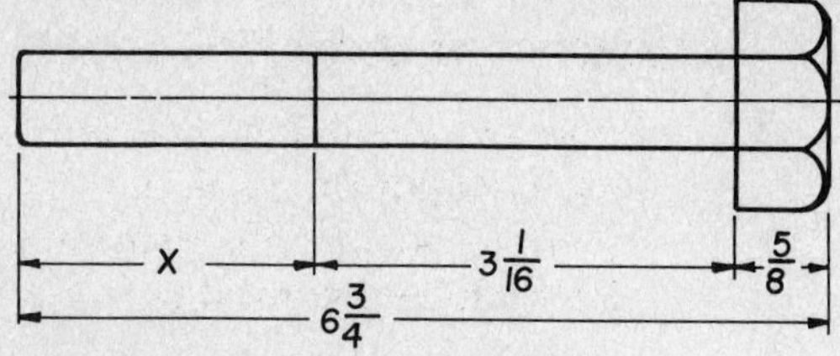

Figure 14-13

APPLICATION PROBLEM 29

Dimension A represents the width of a keyway that is to be milled into a square workpiece in Figure 14-14. What is the width of the keyway?

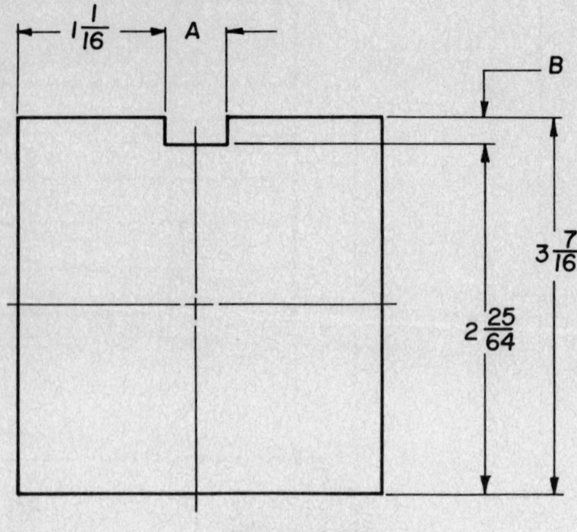

Figure 14-14

APPLICATION PROBLEM 30

Calculate dimension *B* in Figure 14-14.

APPLICATION PROBLEM 31

What is the inside diameter of the cylinder in Figure 14-15?

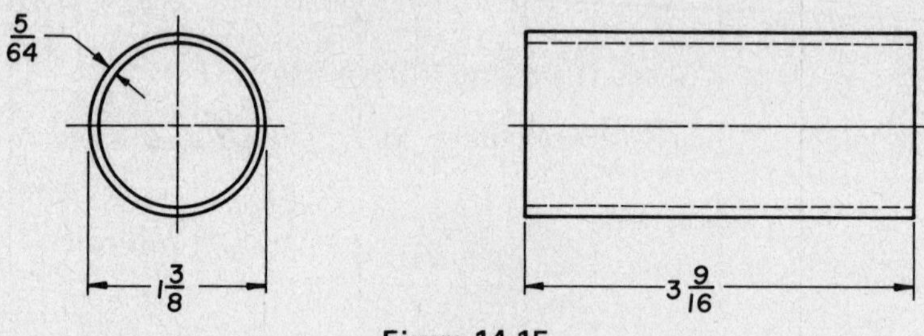

Figure 14-15

APPLICATION PROBLEM 32

You have just machined a "go/no go" gauge that is 5 7/16 in. long. Your supervisor tells you to make another that is twice as long. How long will the new gauge be?

APPLICATION PROBLEM 33

Your job is to make fifteen 3/8-in. nuts from a length of 3/4-in. hexagonal bar. If you must allow 1/32 in. of waste for each nut, how long a piece of hex bar is needed?

APPLICATION PROBLEM 34

At 75 3/4 cents a pound, how much would 256 1/2 pounds of scrap aluminum cost?

APPLICATION PROBLEM 35

If a drop forge can produce 117 plates per hour, how many can it produce in 5 2/3 hours?

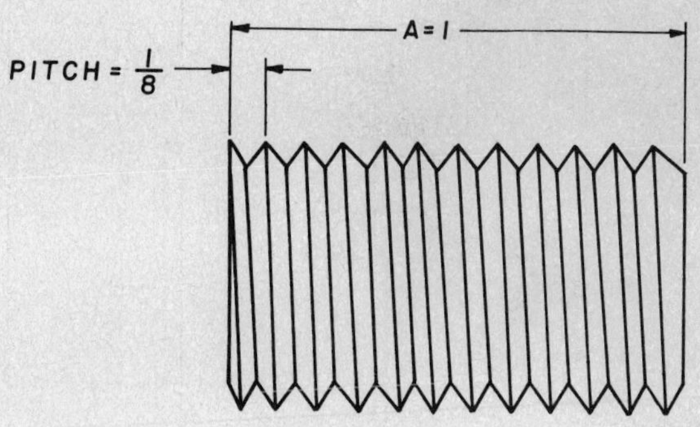

Figure 14-16

APPLICATION PROBLEM 36

You can see from Figure 14-16 that the *pitch* of a thread is the distance from a point on a thread to the corresponding point on another. The *pitch* is usually expressed as a fraction. In this figure, there are eight complete threads. Therefore, the *pitch* is 1 in. divided by eight threads of 1/8 in. If dimension *A* is changed to 3 in. and the total number of threads is 36, what is the *pitch?*

APPLICATION PROBLEM 37

If dimension *A* in Figure 14-16 is changed to 6 in. and the total number of threads is 72, what is the *pitch?*

APPLICATION PROBLEM 38

A can full of washers weighs 17 3/4 pounds. The empty can weighs 1 1/2 pounds. If one of the washers weighs 1/16 of a pound, how many washers does the can hold?

APPLICATION PROBLEM 39

If you must allow 1/16 in. for waste, what length of stock is needed to make the workpiece in Figure 14-17? How many such workpieces could be made from a piece of stock 68 in. long? (Allow 1/16 in. per workpiece for waste.) How much of the stock will be left over?

APPLICATION PROBLEM 40

If you add 1/8 in. to each dimension in Figure 14-17, how many workpieces like it would be made from a piece of stock 54 3/4 in. long? (Allow 1/16 in. per workpiece for waste.) How much of the stock would be left over?

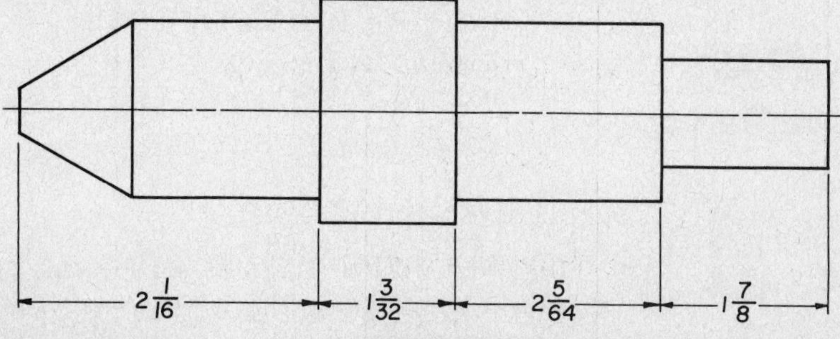

Figure 14-17

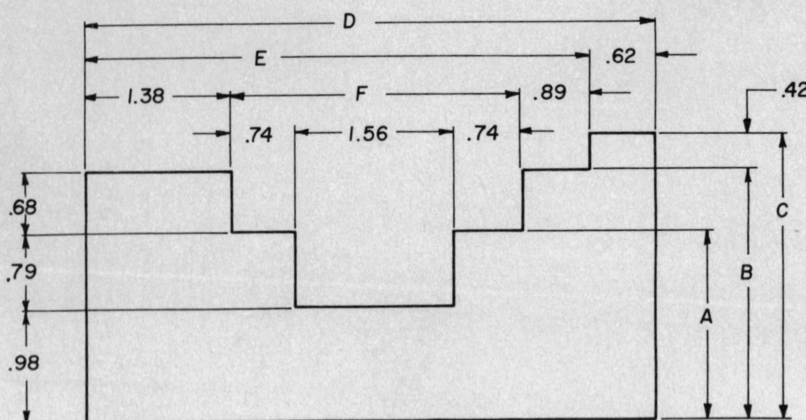

Figure 14-18

APPLICATION PROBLEM 41

Use the information given in Figure 14-18 to calculate dimensions *A*, *B*, *C*, *D*, *E*, and *F*.

APPLICATION PROBLEM 42

Add 0.23 to each of the dimensions given in Figure 14-19. Then calculate dimension *X*.

APPLICATION PROBLEM 43

Refer to Figure 14-19 in solving this problem. Subtract 0.09 in. from all dimensions. Recalculate dimensions *A*, *B*, *C*, *D*, *E*, and *F*.

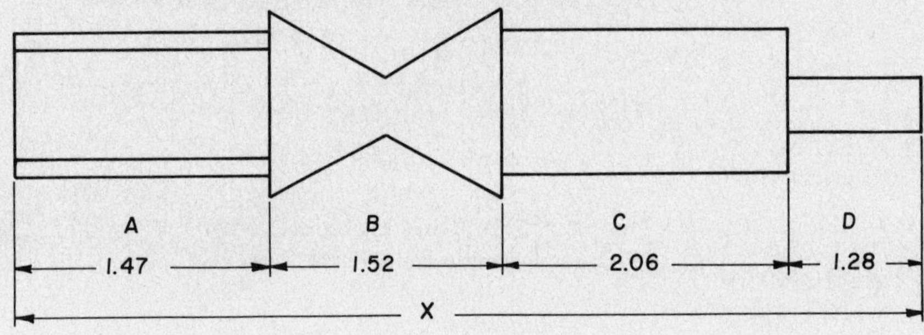

Figure 14-19

APPLICATION PROBLEM 44

Refer to Figure 14-19 in solving this problem. Make the following subtractions and recalculate dimension *X*.

Dimension *A*	Subtract 0.12
Dimension *B*	Subtract 0.013
Dimension *C*	Subtract 0.19
Dimension *D*	Subtract 0.31

APPLICATION PROBLEM 45

Refer to Figure 14-20 in solving this problem. Complete the "total weight" column in the table. Weights given are expressed in ounces. Total weights should also be expressed in ounces.

ITEM NAME	NO. REQ'D	WEIGHT EACH	TOTAL WEIGHT
WASHER	12	4.067	
PIN	94	0.914	
BOLT	37	9.163	
NUT	48	6.539	

Figure 14-20

APPLICATION PROBLEM 46

Add 0.0965 ounce to the weight of each item in Figure 14-20 and recalculate the total weight entry for each item.

APPLICATION PROBLEM 47

Convert the fractions in Figure 14-21 to decimal form. Round each answer to the nearest thousandth.

FRACTION	DECIMAL
$\frac{1}{8}$	
$\frac{3}{16}$	
$\frac{5}{32}$	
$\frac{11}{64}$	
$\frac{17}{128}$	

Figure 14-21

APPLICATION PROBLEM 48

Refer to Figure 14-22 in solving this problem. How many stainless steel sleeves like the one in this figure can be cut from stock 47.09 in. long?

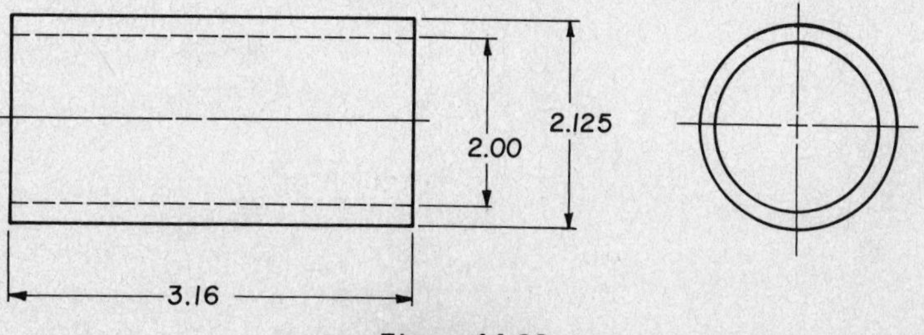

Figure 14-22

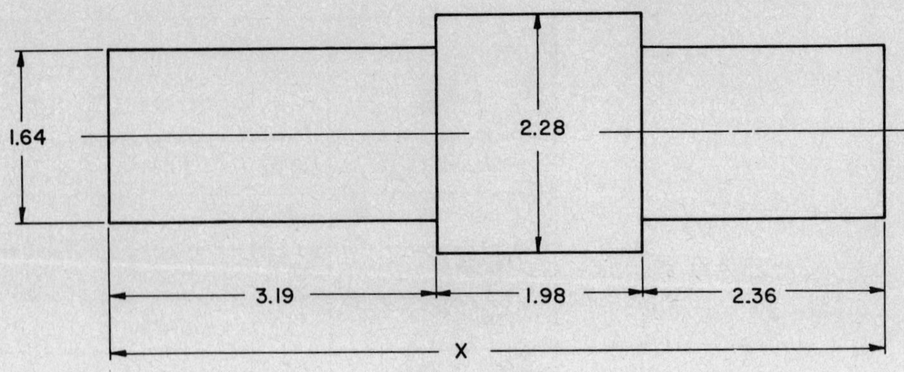

Figure 14-23

APPLICATION PROBLEM 49

Refer to Figure 14-23 in solving this problem.

(a) What is the overall length of the part?
(b) How much would seven such parts weigh in pounds if one of them weighs 0.76 pound?
(c) How many of this part would be made from stock 53.47 in. long?

APPLICATION PROBLEM 50

Refer to Figure 14-23 in solving this problem.

(a) Increase all dimensions by 2.13 times.
(b) How many of the new parts could be made from stock that is 68.76 in. long?

15
APPLYING ALGEBRA AND GEOMETRY IN THE MACHINE TRADES

This is an application chapter. It consists wholly of on-the-job-oriented problems that require you to apply what you have learned in earlier chapters. The problems in this chapter apply what was learned in Chapters 6 to 10.* You are encouraged to use Chapters 6 to 10 and all appended material for reference purposes in solving the problems in this chapter.

APPLICATION PROBLEM 1

Solve the following machine-shop-related equations.

(a) $19.5 + X = 116.03;\ X = ?$

(b) $\dfrac{0.019A}{0.004} + 9 = 12.56;\ A = ?$

APPLICATION PROBLEM 2

Solve the following machine-shop-related equations.

(a) $\dfrac{X + 9}{6} + 8.75 = 156.82;\ X = ?$

(b) $12a + 0.046a = 32.15;\ a = ?$

APPLICATION PROBLEM 3

Solve the following machine-shop-related equations.

(a) $\dfrac{3X + 5X + 2X}{2} = 19.45;\ X = ?$

(b) $\dfrac{0.017b + 1.073}{0.115} = 3.056;\ b = ?$

*Power and root problems (Chapter 5) are also included in this chapter.

APPLICATION PROBLEM 4

Solve the following machine-shop-related equations.

(a) $$\frac{3X + 5X + 9}{3} + \frac{2X - 5.56}{2} = 112;\ X = ?$$

(b) $$\frac{97}{6(9 + 3.56) - 2(3.41 + 2.75)} - 2.61X = 175;\ X = ?$$

Figure 15-1 contains information needed when making thread calculations. Figure 15-2 contains some of the formulas used in making thread calculations for American Standard and Metric threads. Figure 15-3 explains how to interpret thread notes for American Standard and Metric threads. Refer to Figures 15-1, 15-2, and 15-3 in solving Application Problems 5 to 8.

APPLICATION PROBLEM 5

Calculate the following threat values to the nearest thousandth.

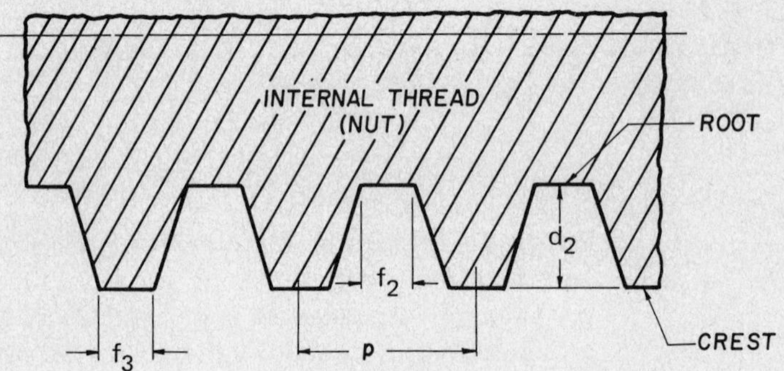

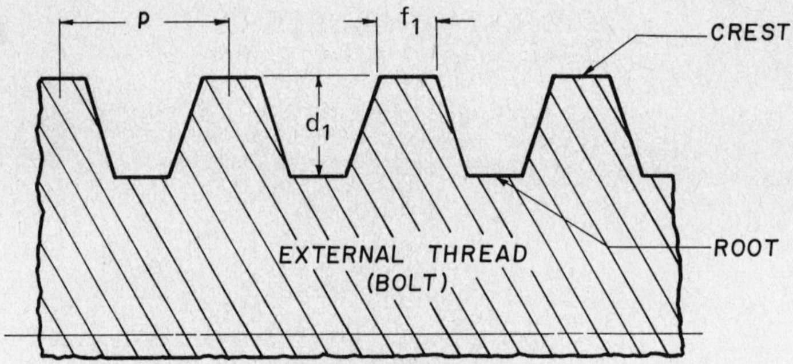

SYMBOL	MEANING
p	PITCH
n	NUMBER OF THREADS PER INCH OR MILLIMETER
f_1	FLAT AT CREST, EXTERNAL CREST
f_2	FLAT AT ROOT, INTERNAL THREAD
f_3	FLAT AT CREST, INTERNAL CREST
d_1	DEPTH, EXTERNAL THREAD
d_2	DEPTH, INTERNAL THREAD

Figure 15-1

TO FIND	AMERICAN STANDARD FORMULA	METRIC FORMULA
PITCH	$p = \frac{1}{n}$	$p = \frac{1}{n}$
NUMBER OF THREADS PER INCH OR MILLIMETER	$n = \frac{1}{p}$	$n = \frac{1}{p}$
FLAT AT CREST, EXTERNAL THREAD	$f_1 = 0.125 \times p$	$f_1 = 0.125 \times p$
FLAT AT ROOT, INTERNAL THREAD	$f_2 = 0.125 \times p$	$f_2 = 0.125 \times p$
FLAT AT CREST, INTERNAL THREAD	$f_3 = 0.250 \times p$	$f_3 = 0.250 \times p$
DEPTH, EXTERNAL THREAD	$d_1 = 0.61343 \times p$	$d_1 = 0.61343 \times p$
DEPTH, INTERNAL THREAD	$d_2 = 0.54127 \times p$	$d_2 = 0.54127 \times p$

Figure 15-2

5/8 in. - 12 UNC - 2A

$p =$
$f_1 =$
$f_2 =$
$d_1 =$
$d_2 =$

APPLICATION PROBLEM 6

Calculate the following thread values to the nearest thousandth.

9/16 in. - 18 UNF - 2B

$p =$
$f_1 =$
$f_2 =$
$d_1 =$
$d_2 =$

APPLICATION PROBLEM 7

Calculate the following thread values to the nearest thousandth.

M 10 × 1.75 - 5H

$p =$
$f_1 =$
$f_2 =$
$d_1 =$
$d_2 =$

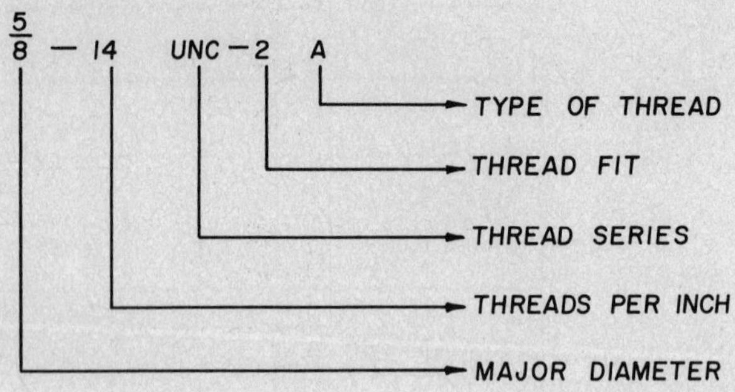

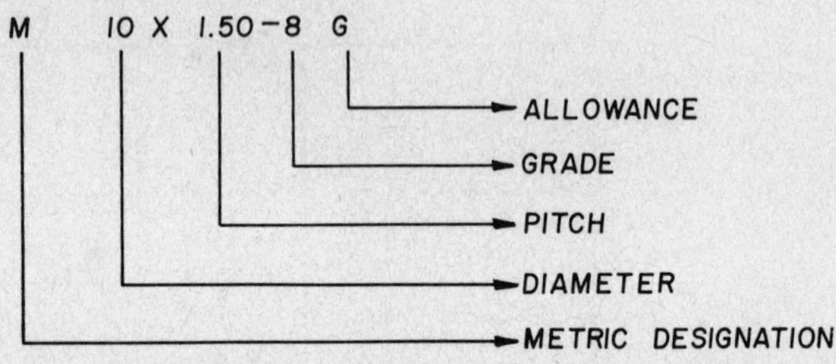

Figure 15-3

APPLICATION PROBLEM 8

Calculate the following thread values to the nearest thousandth.

M 8 × 1.50 - 8g

$p =$

$f_1 =$

$f_2 =$

$d_1 =$

$d_2 =$

Figure 15-4 contains a symbols legend for gear calculation formulas. Figure 15-5 contains the formulas for American Standard and Metric gear calculations. Refer to these figures in solving Application Problems 9 to 12.

APPLICATION PROBLEM 9

Perform the calculations indicated in gear A.

Gear A	
Outside diameter	8.75 in.
Number of teeth	54

$D =$

$P =$

$a =$

$b =$

$c =$

$t =$

SYMBOL	MEANING
D	PITCH DIAMETER
D_o	OUTSIDE DIAMETER
D_R	ROOT DIAMETER
p	DIAMETRAL PITCH
p	CIRCULAR PITCH
t	CIRCULAR THICKNESS
a	APPENDUM
b	DEDENDUM
c	CLEARANCE
h_t	WHOLE DEPTH
h_K	WORKING DEPTH
n	NUMBER OF TEETH
m	MODULE

Figure 15-4

$p =$

$h_t =$

$h_k =$

$D_R =$

APPLICATION PROBLEM 10

Perform the calculations indicated on gear B.

Gear B	
Outside diameter	6.36 in.
Number of teeth	38

$D =$

$P =$

$a =$

$b =$

$c =$

$t =$

$p =$

$h_t =$

$h_k =$

$D_R =$

APPLICATION PROBLEM 11

Perform the calculations indicated on gear C.

Gear C	
Outside diameter	12.75 in.
Number of teeth	78

TO FIND	AMERICAN NATIONAL STANDARD	METRIC
MODULE	. . . .	$m = \frac{D}{n}$
DIAMETRAL PITCH	$p = \frac{n}{D}$	$p = \frac{1}{m}$
PITCH DIAMETER	$D = \frac{n}{p}$ $D = \frac{Do \times n}{(n + 2)}$	$D = n \times m$
NUMBER OF TEETH (EXPRESSED AS A WHOLE NUMBER)	$n = P \times D$	$n = \frac{D}{m}$
ADDENDUM	$a = \frac{1}{P}$	$a = m$
DEDENDUM (preferred)	$b = \frac{1.250}{P}$	$b = 1.250 \times m$
CLEARANCE (preferred)	$c = \frac{0.250}{P}$	$c = 0.250 \times m$
CLEARANCE (minimum)	$c = \frac{0.157}{P}$	$c = 0.157 \times m$
CIRCULAR THICKNESS-BASIC	$t = \frac{1.5708}{P}$	$t = 1.5708 \times m$
ROOT DIAMETER	$Dr = \frac{(n - 2.5)}{P}$ $D_R = D - (2 \times b)$	$D_R = D - (2.5 \times m)$
OUTSIDE DIAMETER	$Do = \frac{(n + 2)}{P}$ $Do = D + (2 \times a)$	$Do = m \times (n + 2)$ $Do = D + (2 \times m)$
WHOLE DEPTH (preferred)	$h_t = \frac{2.250}{P}$ $ht = a + b$	$h_t = a + b$
WORKING DEPTH	$h_k = \frac{2}{P}$ $h_k = a + b - c$	$h_k = 2 \times a$
CIRCULAR PITCH	$p = \frac{3.1416}{P}$	$p = 3.141\,6 \times m$

Figure 15-5

$D =$

$P =$

$a =$

$b =$

$c =$

$t =$

$p =$

$h_t =$

$h_k =$

$D_R =$

APPLICATION PROBLEM 12

Perform the gear calculations indicated on gear D.

Gear D	
Outside diameter	4.68 in.
Number of teeth	24

$D =$

$P =$

$$a =$$
$$b =$$
$$c =$$
$$t =$$
$$p =$$
$$\text{ht} =$$
$$\text{hk} =$$
$$\text{DR} =$$

Figure 15-6 contains a symbols legend for speed and feed calculation formulas. Figure 15-7 contains the formulas for American Standard and Metric units. Refer to these figures in solving Application Problems 13 to 16.

SYMBOL	MEANING	AMERICAN STANDARD UNITS	METRIC UNITS
V	CUTTING SPEED	FEET PER MINUTE (fpm)	METERS PER MINUTE (m/min)
D	DIAMETER	INCHES (in.)	MILLIMETERS (mm)
N	SPINDLE SPEED	REVOLUTIONS PER MINUTE (rpm)	REVOLUTIONS PER MINUTE (r/min)

Figure 15-6

TO FIND	AMERICAN STANDARD UNITS	METRIC UNITS
N	$N = \dfrac{12 \times V}{3.1416 \times D}$	$N = \dfrac{1000 \times V}{3.1416 \times D}$
V	$V = \dfrac{3.1416 \times D \times N}{12}$	$V = \dfrac{3.1416 \times D \times N}{1000}$
D	$D = \dfrac{V \times 12}{3.1416 \times N}$	$D = \dfrac{V \text{ X } 1000}{3.1416\text{X}N}$

Figure 15-7

APPLICATION PROBLEM 13

A piece of 15/16-in.-diameter stainless steel stock is turned in a lathe at 156 revolutions per minute. What is the cutting speed?

APPLICATION PROBLEM 14

A piece of 5/8-in.-diameter iron stock is turned on a lathe. The cutting speed needs to be 12 ft per minute. How many revolutions per minute are required?

APPLICATION PROBLEM 15

How many revolutions per minute are required to turn a 7/8-in.-diameter tool steel rod at 156 ft per minute?

APPLICATION PROBLEM 16

If the cutting speed is 75 ft per minute, how many revolutions per minute should a 3 1/2-in. end mill be run to mill a layway 86 in. wide in a stainless steel bar?

Figure 15-8 contains a symbols legend for taper calculations. Figure 15-9 contains the formulas used for taper calculations. Refer to these figures in solving Application Problems 17 to 20.

SYMBOL	MEANING
tpi	TAPER PER INCH
tpf	TAPER PER FOOT
D	DIAMETER, LARGER END
d	DIAMETER, SMALL END
L	LENGTH (IN INCHES)

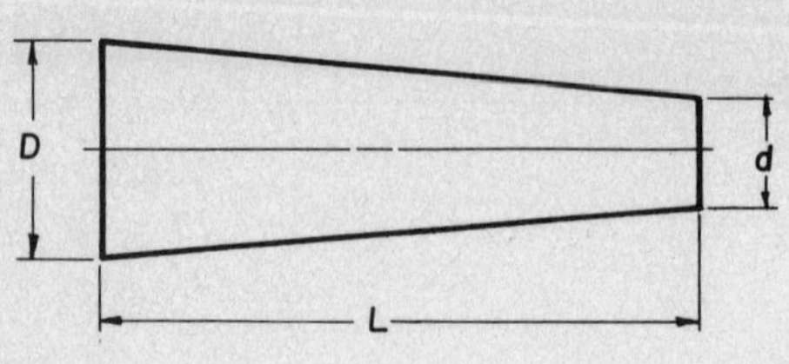

Figure 15-8

TO FIND	KNOWN	FORMULA
tpf	tpi	$tpf = tpi \times 12$
	D,d,L	$tpf = 12\left(\frac{D - d}{L}\right)$
tpi	tpf	$tpi = \frac{tpf}{12}$
	D,d,L	$tpi = \frac{D - d}{L}$
d	D,L,tpf	$d = D - \left[L\left(\frac{tpf}{12}\right)\right]$
D	d,L,tpf	$D = d + \left[L\left(\frac{tpf}{12}\right)\right]$
L	D,d,tpf	$L = 12\left(\frac{D - d}{tpf}\right)$

Figure 15-9

APPLICATION PROBLEM 17

A taper is 9.46 in. long. If $D = 3.96$ in. and $d = 1.75$, what is the taper per foot?

APPLICATION PROBLEM 18

A taper is 5.631 in. long. $D = 1.78$ and the taper is 1.12 in. per foot. What is d?

APPLICATION PROBLEM 19

What is the length of a taper in which $D = 3.87$ in., $d = 1.08$ in., and the taper is 0.68 in. per foot?

APPLICATION PROBLEM 20

Calculate the total amount of taper in a part that is 8.75 in. long and has a 0.56 in. per foot taper.

APPLICATION PROBLEM 21

Calculate the ratios of the following gears.

Gear X has 30 teeth.
Gear Y has 10 teeth.

(a) What is the ratio of gear X to gear Y?
(b) What is the ratio of gear Y to gear X?

APPLICATION PROBLEM 22

A condensed pickling solvent must be dissolved before it is used. When used it should be a mix of 15 parts solvent and 5 parts water.

(a) What is the ratio of solvent to water?
(b) In mixing the solution, if you use 3 quarts of solvent, how many quarts of water should you use?

APPLICATION PROBLEM 23

Gear A and gear B have a 5:2 ratio. If gear A has 20 teeth, how many teeth does gear B have?

APPLICATION PROBLEM 24

Gear A and gear B have a 4:3 ratio. If gear B has 6 teeth, how many teeth does gear A have?

APPLICATION PROBLEM 25

A machinist is able to produce 15 parts every 70 minutes. How long should it take him to produce 22 parts?

APPLICATION PROBLEM 26

The monthly output of a shop employing 156 machinists is 372 jobs. How many more machinists will be needed to increase the monthly output to 456 jobs?

APPLICATION PROBLEM 27

A machinist can mill 9 keyways per hour. How long (expressed in hours and minutes) will it take to mill 14 keyways?

Figure 15-10 contains an illustration of a series of pulleys. Use this figure for Application Problems 28 to 30.

APPLICATION PROBLEM 28

Pulley 1 is 15.62 in. in diameter and turns at a rate of 172 revolutions per minute. Pulley 2 is 3.78 in. in diameter. Pulley 3 is 12.15 in. in diameter. What size must pulley 4 be to produce 988 revolutions in pulley 4 alone?

APPLICATION PROBLEM 29

Pulley 4 is 2.78 in. in diameter and turns at a rate of 1200 revolutions per minute. Pulley 3 is 8.42 in. in diameter. Pulley 2 is 4.26 in. in diameter. What size is pulley 1 if it turns at 213 revolutions per minute?

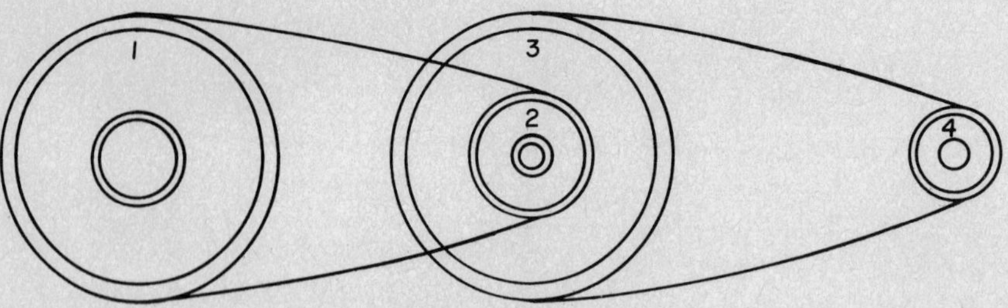

Figure 15-10

APPLICATION PROBLEM 30

Pulley 1 is 26.52 in. in diameter. Pulley 3 is 37.89 in. in diameter. Pulley 4 is 9.76 in. in diameter. If the rate of pulley 1 is 147 revolutions per minute and the rate of pulley 4 is 1036 revolutions per minute, what is the diameter of pulley 2?

APPLICATION PROBLEM 31

What is 0.4572 raised to the third power? The fourth power?

APPLICATION PROBLEM 32

What is 1.57 raised to the second power? The third power?

APPLICATION PROBLEM 33

Find the value of each of the following expressions.

(a) $X^{12} \div X^7$
(b) $A^5 \times A^2 \times A^4$

APPLICATION PROBLEM 34

Find the value of each of the following expressions.

(a) $\dfrac{F^4 \times F^2}{F}$

(b) $\dfrac{7^2 + 5^3}{2^3}$

APPLICATION PROBLEM 35

Find the value of each of the following expressions.

(a) $\dfrac{X^2 \times X^3 \times X^4}{X^5}$

(b) $\dfrac{3^2 + 4^3 + 7.89^3}{5.136^3}$

APPLICATION PROBLEM 36

Calculate the square root of each of the following numbers.

(a) 25

(b) 36
(c) 49

APPLICATION PROBLEM 37

Calculate the square root of each of the following numbers.

(a) 172
(b) 1,480
(c) 12,116

APPLICATION PROBLEM 38

Calculate the square root of each of the following numbers.

(a) 21.789
(b) 147.003
(c) 81,917.561

Figure 15-11 contains a generalized example of a drilled plate. Drilling equally spaced holes in such plates is a frequently required task of machinists.

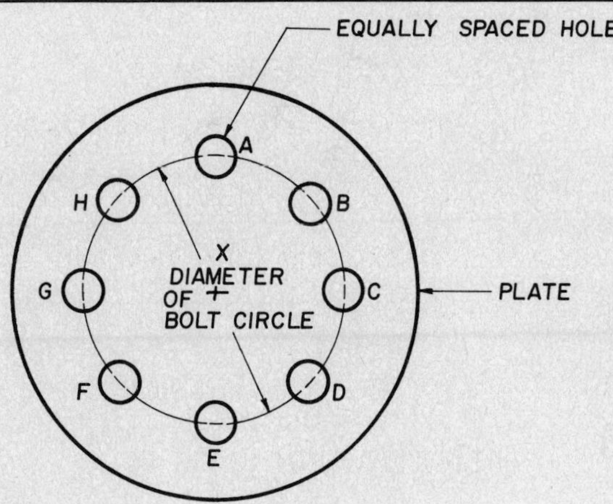

Figure 15-11

APPLICATION PROBLEM 39

Assume the following information with regard to the plate in Figure 15-11.

Plate diameter	12.75 in.
Hole diameter	0.75 in. (all holes)
Number of holes	8

(a) What is the area of the plate's top surface before the holes are drilled?
(b) What is the area of the plate's top surface minus the holes?

APPLICATION PROBLEM 40

If the diameter of the plate in Figure 15-11 is 13.68 in., what is the circumference of the plate?

APPLICATION PROBLEM 41

Assume the following information with regard to the plate in Figure 15-11.

Plate diameter	15.06 in.
Number of holes	4 (*A, C, E,* and *G*)
Hole diameter	0.56 in. (all holes)
Diameter of bolt circle	13.06 in.

(a) What is the area of the plate's top surface minus the holes?
(b) What is the length of arc $\widehat{AC}$?

APPLICATION PROBLEM 42

Assume the following information with regard to Figure 15-11.

Diameter of bolt circle	10.85 in.
Number of holes	8

(a) What is the length of arc $\widehat{BC}$?
(b) What is the area of sector *XAB*?

APPLICATION PROBLEM 43

Assume the following information with regard to Figure 15-11.

Plate diameter	8.68 in.
Diameter of bolt circle	7.18 in.
Number of holes	4 (*B, D, F,* and *H*)

(a) What is the area of the plate?
(b) What is the area of the plate minus sector *XHB*?

APPLICATION PROBLEM 44

Assume the following information with regard to Figure 15-11.

Plate diameter	9.44 in.
Diameter of bolt circle	7.44 in.
Number of holes	8
Diameter of holes	0.68 in.

(a) What is the circumference of the plate?
(b) What is the area of the top surface of the plate minus the holes?
(c) What is the area of sector *XCF*?

APPLICATION PROBLEM 45

Machine trades personnel frequently use electronic calculations. This means that the data they work with must be expressed in digital form. Convert the following angles to digital form.

(a) $116°22'$
(b) $89°14'$

(c) 56°16′34″
(d) 186°19′03″
(e) 341°01′59″

APPLICATION PROBLEM 46

Calculate the missing angles in Figure 15-12A, B, and C.

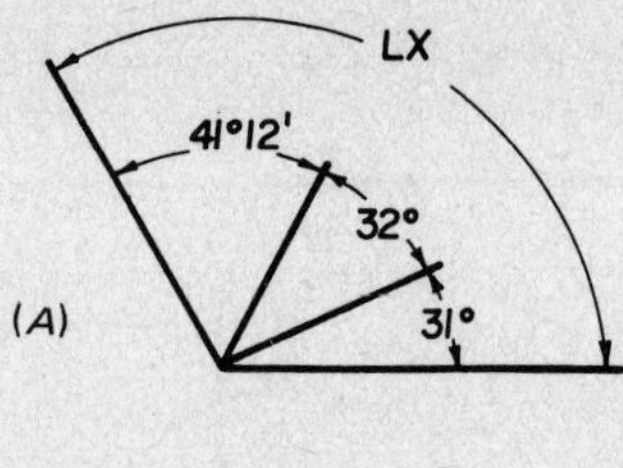

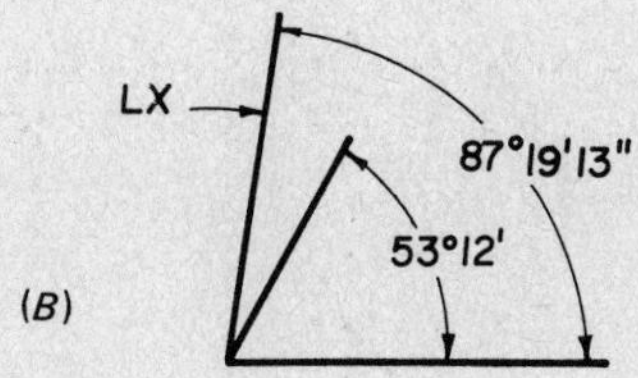

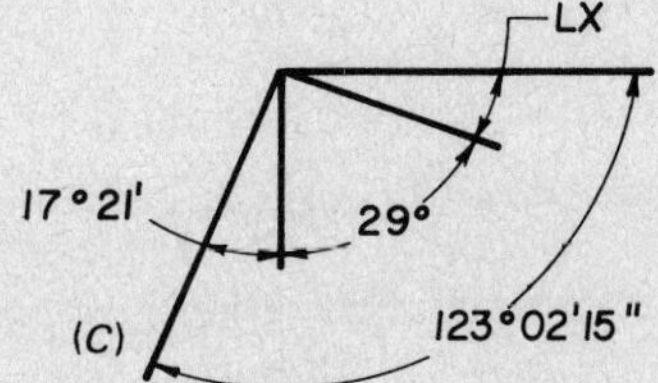

Figure 15-12

APPLICATION PROBLEM 47

The plate in Figure 15-13A is to be cut out on a shearing machine. Calculate the missing angle.

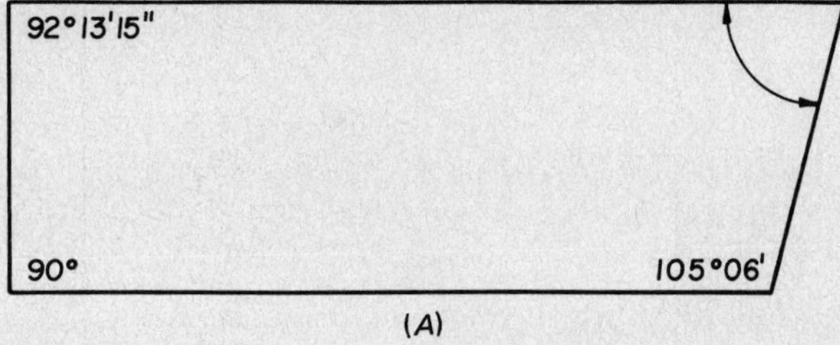

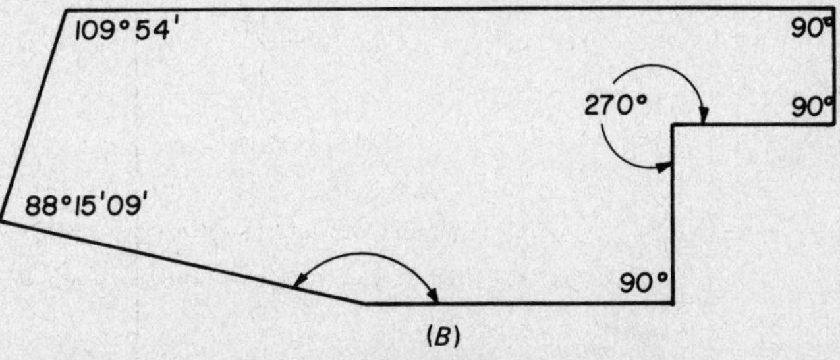

Figure 15-13

APPLICATION PROBLEM 48

The plate in Figure 15-13B is to be cut out on a saw. Calculate the missing angle.

APPLICATION PROBLEM 49

The cover plate in Figure 15-14A is to be cut into seven equal pieces as shown. Calculate angle *X*.

APPLICATION PROBLEM 50

The cover plate in Figure 15-14B is to be cut into five pieces. Pieces b, c, d, and e are equal. Calculate angles *b*, *c*, *d*, and *e*.

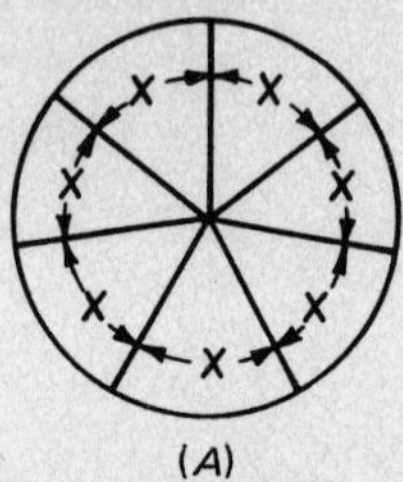

(*A*)

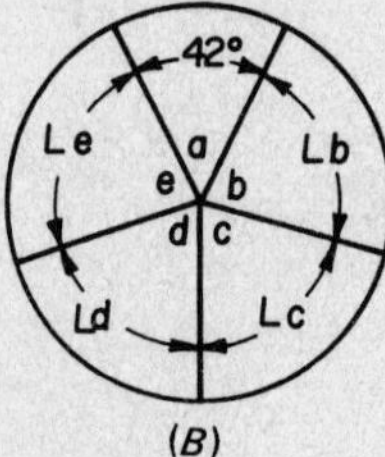

(*B*)

Figure 15-14

16

APPLYING TRIGONOMETRY IN THE MACHINE TRADES

This is an application chapter. It consists solely of on-the-job-oriented problems that require you to apply what you have learned in earlier chapters. The problems in this chapter apply to what was learned in Chapters 11 to 13. You are encouraged to use Chapters 11 to 13 and all appended material for reference purposes in solving the problems in this chapter.

APPLICATION PROBLEM 1

Complete the following chart of natural trigonometric functions.

Angle	*Sine*	*Cosine*	*Tangent*
14°15′	________	________	________
39°17′15″	________	________	________
132°09′	________	________	________
114°59′42″	________	________	________
183°00′17″	________	________	________

APPLICATION PROBLEM 2

Using the following information, find the angles.

	Function	*Angle*
sine	angle $X = 0.8304$	angle $X =$ ________
cosine	angle $Y = 0.9440$	angle $Y =$ ________
tangent	angle $Z = 1.0905$	angle $Z =$ ________

APPLICATION PROBLEM 3

Complete the following chart of trigonometric functions and angles.

Function	*Angle*
cosine = 0.9735	______
tangent = 0.2009	______
sine = 0.8341	______

APPLICATION PROBLEM 4

Complete the following chart of trigonometric functions and angles.

Function	*Angle*
cosine = ______	113°19′57″
tangent = ______	15°01′03″
sine = ______	09°58′47″

Refer to Figure 16-1 for Application Problems 5 to 7.

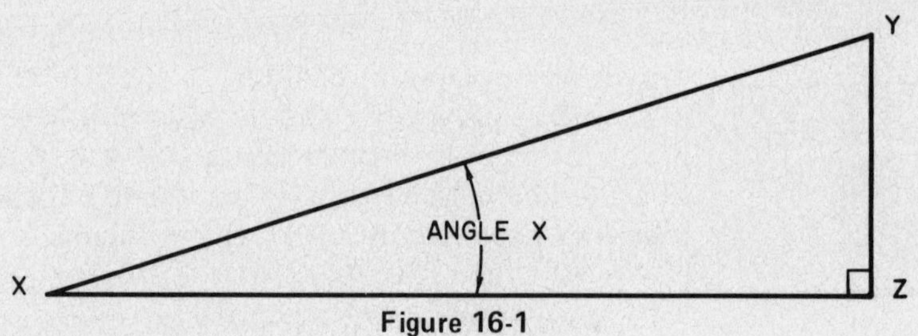

Figure 16-1

APPLICATION PROBLEM 5

Use the following information to solve triangle XYZ.

angle $X = 37°15'$
side $XZ = 72.15$ in.
side $YZ = ?$
side $XY = ?$
angle $Y = ?$

APPLICATION PROBLEM 6

Use the following information to solve triangle XYZ.

angle $Y = 47°09'15''$
side $XY = 15.06$ in.
angle $X = ?$
side $XZ = ?$
side $YZ = ?$

APPLICATION PROBLEM 7

Use the following information to solve triangle XYZ.

side $XZ = 88.55$ in.
angle $X = 12°59'58''$
side $YZ = ?$
side $XY = ?$
angle $Y = ?$

Refer to Figure 16-2 for Application Problems 8 to 13. This figure shows the configuration for a series of plates that are to be drilled. Each plate will receive three holes. You must use the following information to solve for the missing a and b dimensions. These dimensions will be used for positioning the drill table.

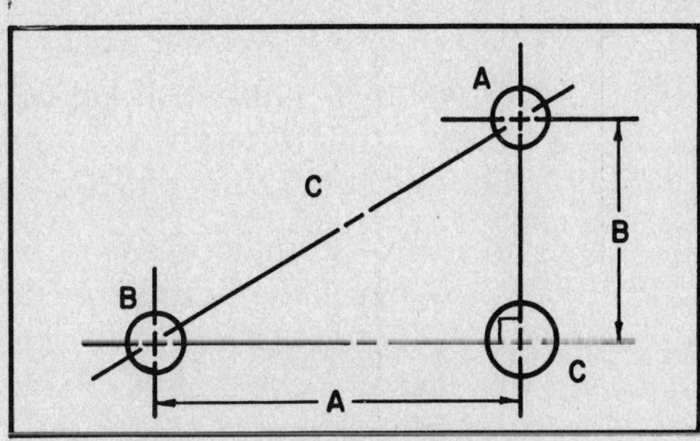

Figure 16-2

APPLICATION PROBLEM 8

Solve for sides a and b if angle $b = 36°15'$ and side $c = 9.431$ in.

APPLICATION PROBLEM 9

Solve for sides a and b if angle $b = 44°13'59''$ and side $c = 15.92$ in.

APPLICATION PROBLEM 10

Solve for sides a and b if angle $b = 86°19'44''$ and side $c = 72.01$ in.

APPLICATION PROBLEM 11

Solve for sides a and b if angle $a = 67°14'$ and side $c = 6.17$ in.

APPLICATION PROBLEM 12

Solve for sides a and b if angle $a = 12°12'09''$ and side $c = 92.075$ in.

APPLICATION PROBLEM 13

Solve for sides a and b if angle $a = 34°28'14''$ and side $c = 56.251$ in.

Refer to Figure 16-3 for Application Problems 14 and 15. It represents the configuration for six holes that are to be drilled into a series of plates.

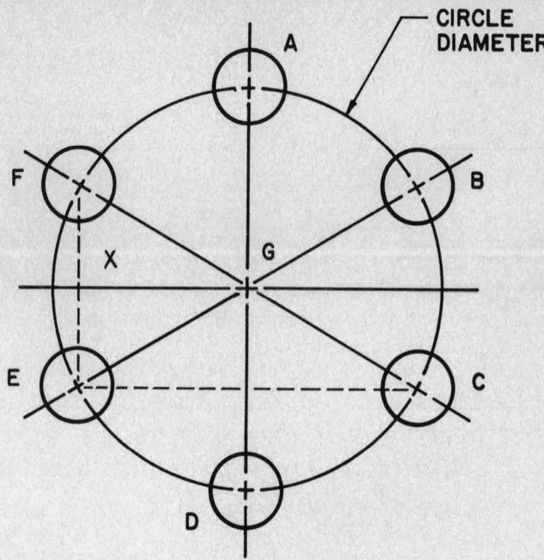

Figure 16-3

APPLICATION PROBLEM 14

A machinist must drill six equally spaced holes configured like those in Figure 16-3 through a 0.50-in.-thick stainless steel plate. If the circle diameter is 6.72 in., calculate the following.

(a) What is angle *FGX*?
(b) What is the length of side *FG*?
(c) What is the length of side *GX*?
(d) What is the length of side *FX*?
(e) What is angle *GFX*?
(f) What is the length of $\overline{EC}$?

APPLICATION PROBLEM 15

Change the circle diameter in Figure 16-3 to 12.73 in. and redo Application Problem 14.

Machine trade personnel frequently use trigonometry to calculate missing dimensions when drilling holes in workpieces. In Application Problems 16 to 20, calculate the missing dimensions.

APPLICATION PROBLEM 16

Calculate the missing dimensions and angles in Figure 16-4.

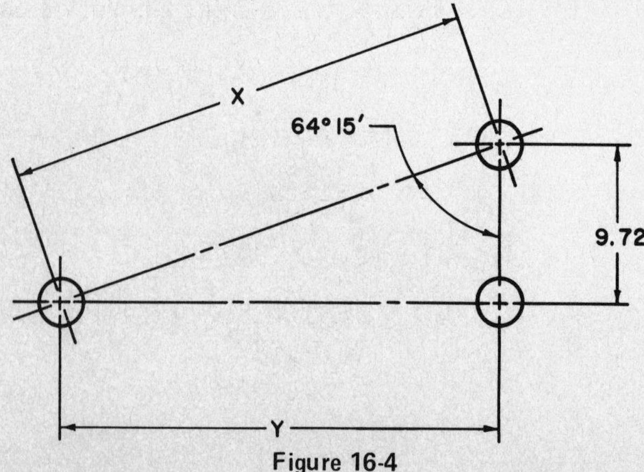

Figure 16-4

APPLICATION PROBLEM 17

Calculate the missing dimensions and angles in Figure 16-5.

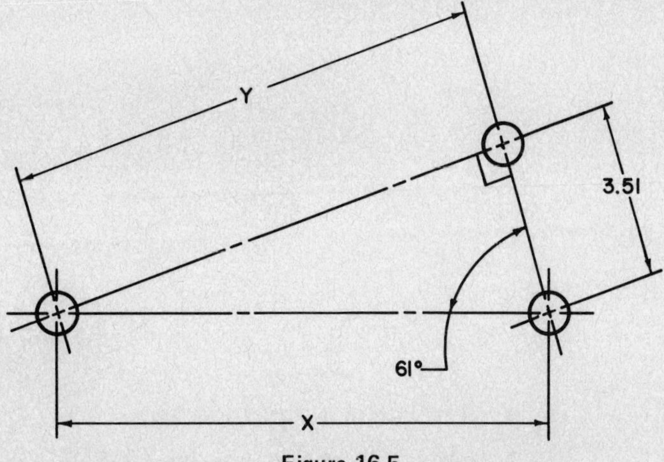

Figure 16-5

APPLICATION PROBLEM 18

Calculate the missing dimensions and angles in Figure 16-6.

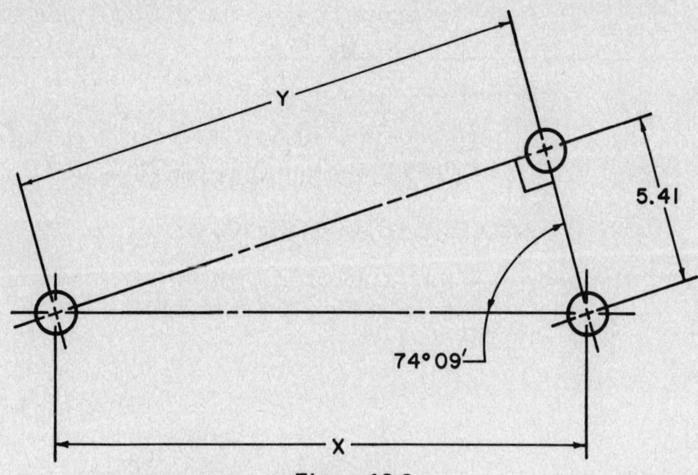

Figure 16-6

APPLICATION PROBLEM 19

Calculate the missing dimensions and angles in Figure 16-7. (Assume that $Z = 2.05$ in.)

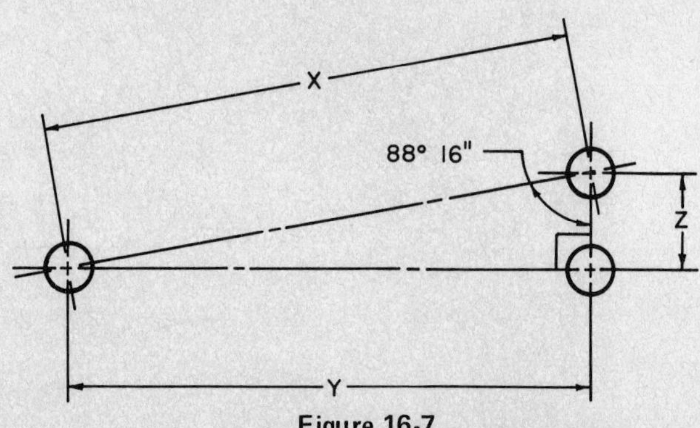

Figure 16-7

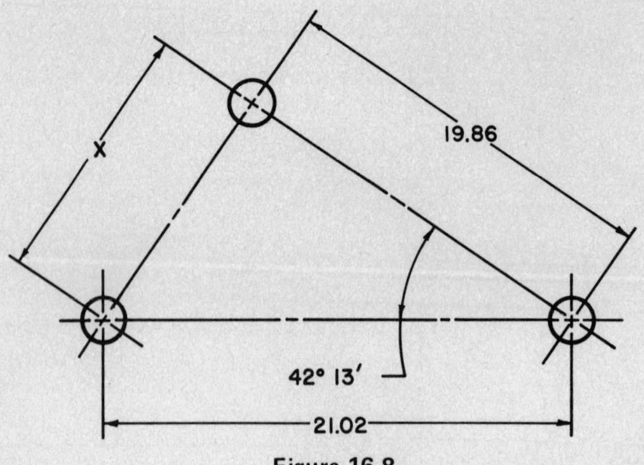

Figure 16-8

APPLICATION PROBLEM 20

Calculate the missing dimensions and angles in Figure 16-8.

Sine bars are used in the machine shop for measuring angles that have been cut in parts. Sine bar calculations are common in the machine trades. Refer to Figure 16-9 in completing Application Problems 21 to 24.

APPLICATION PROBLEM 21

If the sine bar in Figure 16-9 is 14.67 in. long and the gauge block height is 4.30 in., what is the angle?

APPLICATION PROBLEM 22

If the sine bar in Figure 16-9 is 9.89 in. long and the gauge block height is 6.38 in., what is the angle?

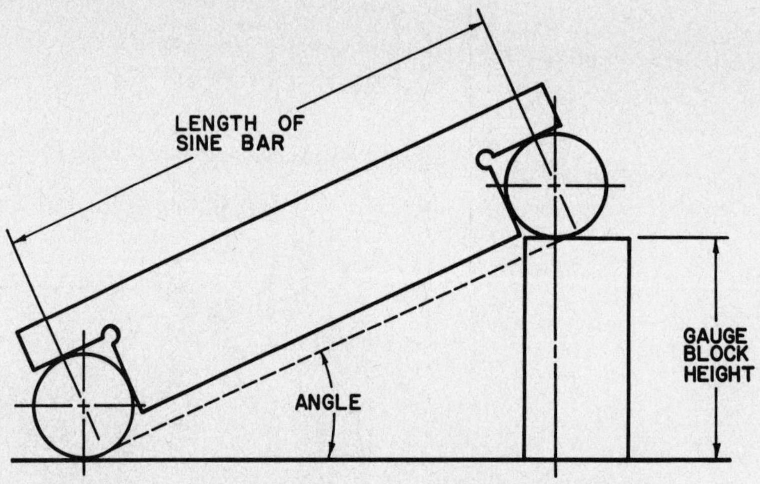

$$\text{SINE OF ANGLE} = \frac{\text{GAUGE BLOCK HEIGHT}}{\text{LENGTH OF SINE BAR}}$$

Figure 16-9

APPLICATION PROBLEM 23

If the angle in Figure 16-9 is 17°14′16″ and the gauge block height is 4.89 in., what is the length of the sine bar?

APPLICATION PROBLEM 24

If the angle in Figure 16-9 is 32°09′15″ and the sine bar is 12.98 in. long, what is the height of the gauge block?

Tape and bevel problems are common in the machine trades. Often they can be solved by dividing the workpiece (on paper) into triangles and rectangles. Use this method for solving Application Problems 25 to 28.

APPLICATION PROBLEM 25

What is the total angle of taper in Figure 16-10?

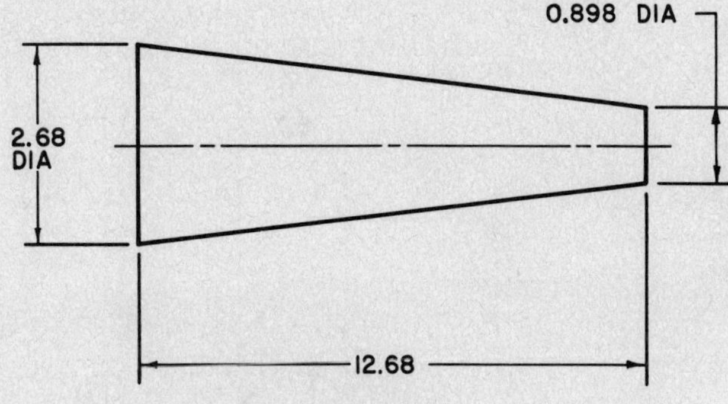

Figure 16-10

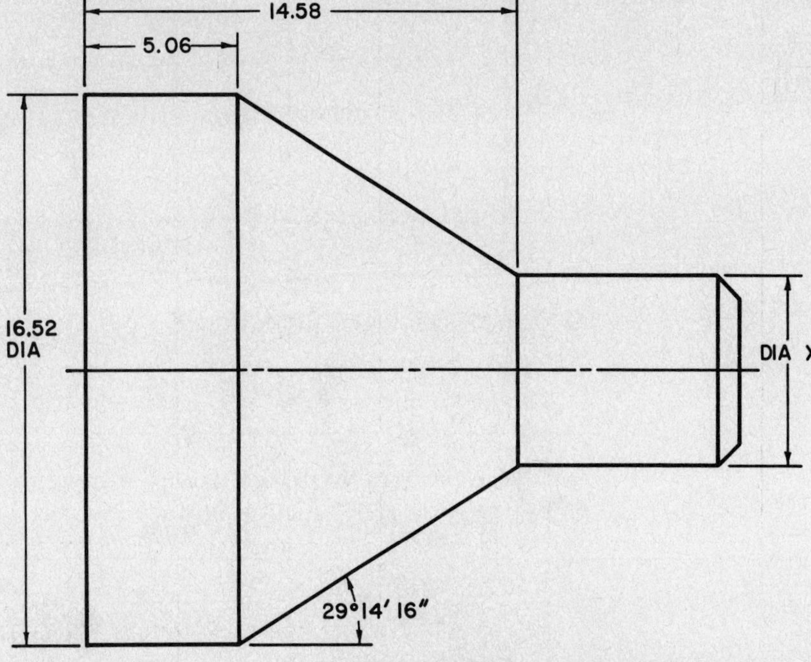

Figure 16-11

APPLICATION PROBLEM 26

What is the total angle of taper in Figure 16-10 if the large diameter is increased by 0.75 in.?

APPLICATION PROBLEM 27

Calculate the missing diameter X in Figure 16-11.

APPLICATION PROBLEM 28

Calculate the missing diameter X in Figure 16-11 if the angle is changed to 36°19′24″.

Machine trade personnel are often required to calculate cutting depths. Use triangles to calculate the cutting depths in Application Problems 29 and 30.

APPLICATION PROBLEM 29

What is the depth of the V cut into the block in Figure 16-12?

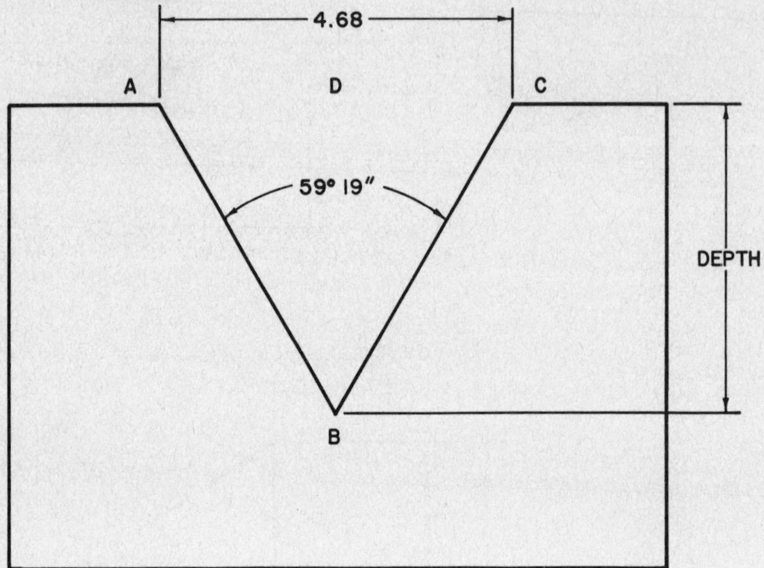

Figure 16-12

APPLICATION PROBLEM 30

What is the depth of the V cut into the block in Figure 16-12 if the angle is changed to 64°32′ and the width is changed to 5.49 in.?

Machine trade personnel are often required to do dovetail calculations. Refer to Figure 16-13 for Application Problems 31 to 33.

APPLICATION PROBLEM 31

In Figure 16-13, line segment *ab* bisects the 68°14″ angle. Calculate dimensions Y and Z.

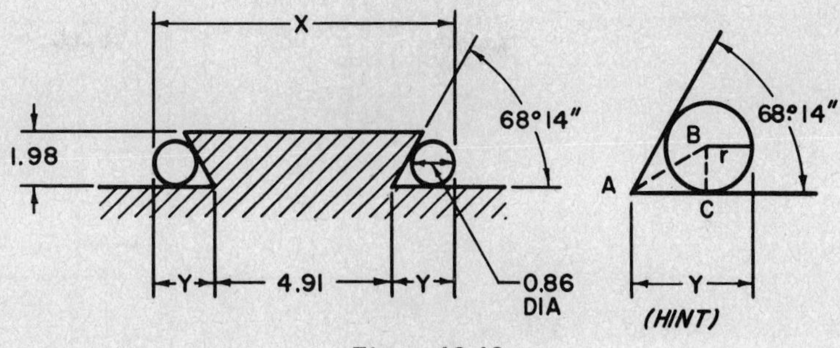

Figure 16-13

APPLICATION PROBLEM 32

If the angle in Figure 16-13 is changed to 73°19′ and the diameter dimension is changed to 0.975 in., what are dimensions Y and X?

APPLICATION PROBLEM 33

If the diameter dimension in Figure 16-13 is changed to 0.672 in. and dimension Y is 1.875 in., what is the angle?

Machine trades personnel are often required to calculate the area of work surfaces. Calculate the areas of the surfaces in Application Problems 34 to 39.

APPLICATION PROBLEM 34

The top surface of the triangular workpiece in Figure 16-14 is to be painted. How much surface area is to be painted?

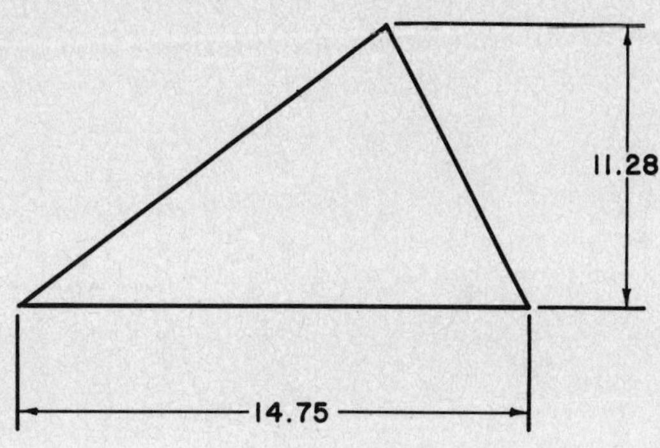

Figure 16-14

APPLICATION PROBLEM 35

How much surface area must be painted if the dimensions in Figure 16-14 are increased by 1.34 in.?

APPLICATION PROBLEM 36

Figure 16-15 shows the configuration for a triangular plate that is to be machined. The plate contains a rectangular hole. Use the following information to calculate the remaining surface area of the plate.

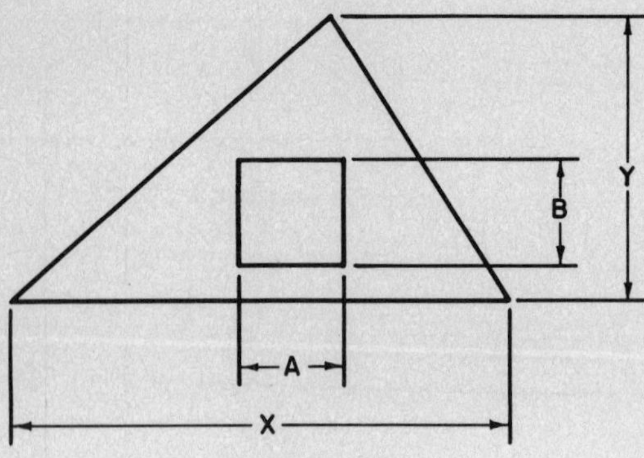

Figure 16-15

$X = 17.62$ in.
$Y = 14.56$ in.
$a = 2.37$ in.
$b = 3.39$ in.

APPLICATION PROBLEM 37

Refer to Figure 16-15 and use the following information to calculate the remaining surface area of the workpiece.

$X = 15.32$ in.
$Y = 12.96$ in.
$a = 1.98$ in.
$b = 1.02$ in.

APPLICATION PROBLEM 38

Refer to Figure 16-16 in solving this problem. What is the area of the workpiece?

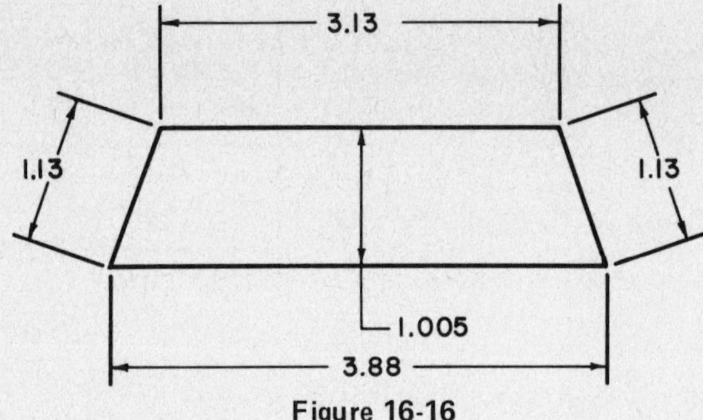

Figure 16-16

APPLICATION PROBLEM 39

Increase each dimension in Figure 16-16 by 0.27 in. What is the area of the workpiece?

APPLICATION PROBLEM 40

Refer to Figure 16-17 in solving this problem. What is the area of the workpiece with the hole removed?

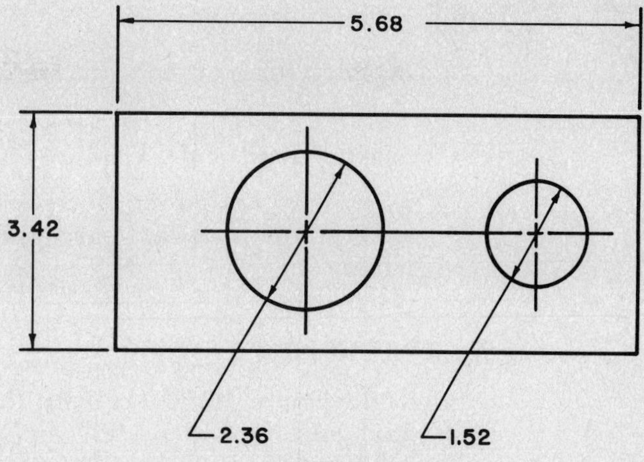

Figure 16-17

Application Problems 41 to 50 may require you to use any or all of the principles explained in Chapters 8 to 13.

APPLICATION PROBLEM 41

Refer to Figure 16-18 in solving this problem. Calculate angle X in this figure.

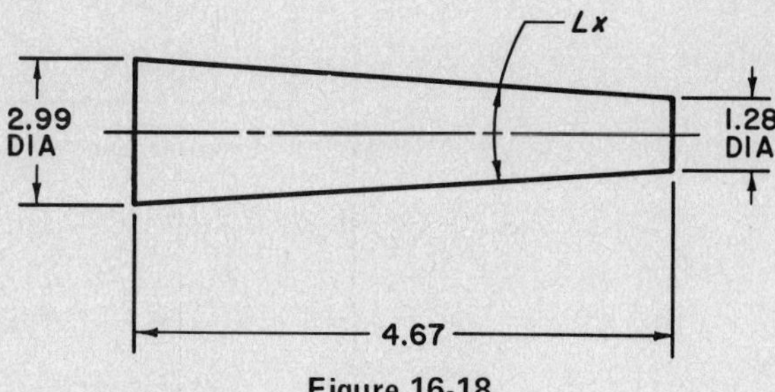

Figure 16-18

APPLICATION PROBLEM 42

Calculate the area of the workpiece in Figure 16-18.

APPLICATION PROBLEM 43

Refer to Figure 16-19 in solving this problem. Calculate the missing dimension X in this figure.

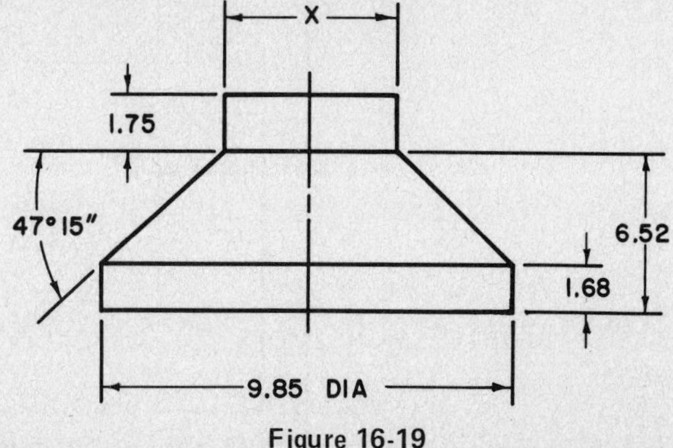

Figure 16-19

APPLICATION PROBLEM 44

Calculate the area of the workpiece in Figure 16-19.

APPLICATION PROBLEM 45

A design error caused the workpiece in Figure 16-19 to be rejected. It must be produced again. Change the angle to 58°17′14″ and recalculate dimension *X*.

APPLICATION PROBLEM 46

Refer to Figure 16-20 in solving this problem. Calculate dimension *X* in this figure.

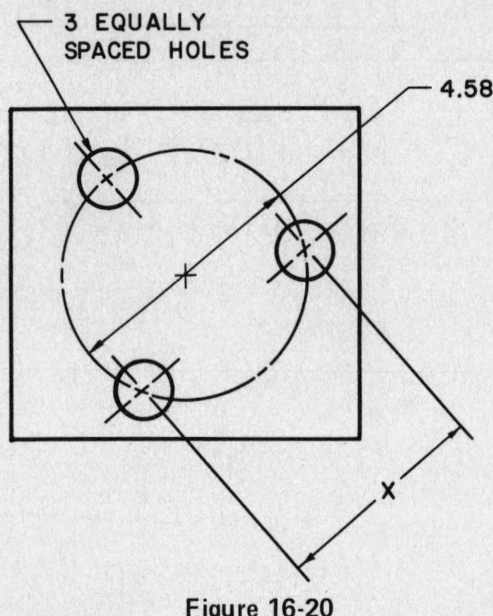

Figure 16-20

APPLICATION PROBLEM 47

The plate in Figure 16-20 is 6 in. square. The holes are 0.75 in. in diameter. What is the area of the plate with the holes removed?

APPLICATION PROBLEM 48

Refer to Figure 16-21 in solving this problem. Calculate the depth of the cut (dimension *C*) if dimension *B* is 6.91 in.

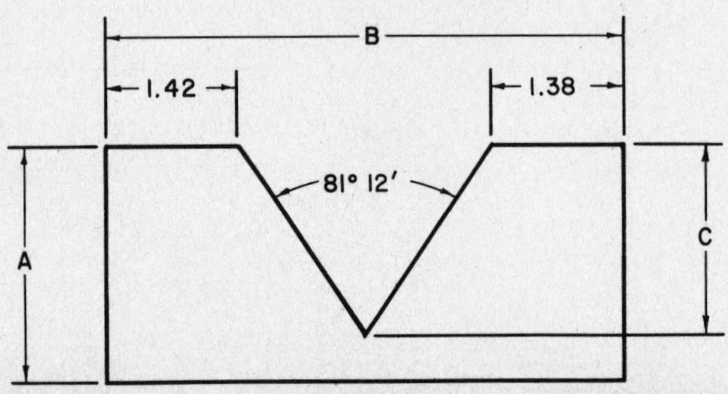

Figure 16-21

APPLICATION PROBLEM 49

Calculate the depth of cut (dimension C) in Figure 16-21 if dimension B is 5.87 in. and the cut angle is $68°12'14''$.

APPLICATION PROBLEM 50

Based on your calculations in Application Problem 49, what is the area of the workpiece?

Appendix

A

TABLES

Learners are encouraged to make full use of this appendix. All materials contained herein have been selected because of their frequent use on the job. The authors surveyed the charts, tables, and reference notes most frequently used by technicians in selecting material for this appendix. Consequently, everything contained herein has value.

Rather than attempting to memorize formulas and equations, learners may find it beneficial to use this appendix as ready reference for these things. The materials contained herein should be particularly helpful to learners attempting to solve the Application Projects in Part 3.

TABLE A-1

Decimal and Metric Equivalents of Fractions of an Inch

Fraction	$\frac{1}{32}$*ds*	$\frac{1}{64}$*ths*	*Decimal*	*Millimeters*	*Fraction*	$\frac{1}{32}$*ds*	$\frac{1}{64}$*ths*	*Decimal*	*Millimeters*
		1	0.015625	0.3968			33	0.515625	13.0966
	1	2	0.03125	0.7937		17	34	0.53125	13.4934
		3	0.046875	1.1906			35	0.546875	13.8903
$\frac{1}{16}$	2	4	0.0625	1.5875	$\frac{9}{16}$	18	36	0.5625	14.2872
		5	0.078125	1.9843			37	0.578125	14.6841
	3	6	0.09375	2.3812		19	38	0.59375	15.0809
		7	0.109375	2.7780			39	0.609375	15.4778
$\frac{1}{8}$	4	8	0.125	3.1749	$\frac{5}{8}$	20	40	0.625	15.8747
		9	0.140625	3.5718			41	0.640625	16.2715
	5	10	0.15625	3.9686		21	42	0.65625	16.6684
		11	0.171875	4.3655			43	0.671875	17.0653
$\frac{3}{16}$	6	12	0.1875	4.7624	$\frac{11}{16}$	22	44	0.6875	17.4621
		13	0.203125	5.1592			45	0.703125	17.8590
	7	14	0.21875	5.5561		23	46	0.71875	18.2559
		15	0.234375	5.9530			47	0.734375	18.6527
$\frac{1}{4}$	8	16	0.25	6.3498	$\frac{3}{4}$	24	48	0.75	19.0496
		17	0.265625	6.7467			49	0.765625	19.4465
	9	18	0.28125	7.1436		25	50	0.78125	19.8433
		19	0.296875	7.5404			51	0.796875	20.2402
$\frac{5}{16}$	10	20	0.3125	7.9373	$\frac{13}{16}$	26	52	0.8125	20.6371
		21	0.328125	8.3342			53	0.828125	21.0339
	11	22	0.34375	8.7310		27	54	0.84375	21.4308
		23	0.359375	9.1279			55	0.859375	21.8277
$\frac{3}{8}$	12	24	0.375	9.5248	$\frac{7}{8}$	28	56	0.875	22.2245
		25	0.390625	9.9216			57	0.890625	22.6214
	13	26	0.40625	10.3185		29	58	0.90625	23.0183
		27	0.421875	10.7154			59	0.921875	23.4151
$\frac{7}{16}$	14	28	0.4375	11.1122	$\frac{15}{16}$	30	60	0.9375	23.8120
		29	0.453125	11.5091			61	0.953125	24.2089
	15	30	0.46875	11.9060		31	62	0.96875	24.6057
		31	0.484375	12.3029			63	0.984375	25.0026
$\frac{1}{2}$	16	32	0.5	12.6997	1	32	64	1.	25.3995

TABLE A-2
Review of Algebra and Geometry

Algebraic Review

$\frac{a}{d} + \frac{b}{d} = \frac{a+b}{d}$ $\frac{a}{b} + \frac{c}{d} = \frac{ad+bc}{bd}$ $\frac{a}{b} \cdot \frac{c}{d} = \frac{ac}{bd}$ $\frac{a/b}{c/d} = \frac{ad}{bc}$

$ax + ay = a(x+y)$ $x^2 - y^2 = (x+y)(x-y)$

$x^2 + 2xy + y^2 = (x+y)^2$ $x^2 - 2xy + y^2 = (x-y)^2$

$x^2 + (a+b)x + ab = (x+a)(x+b)$

$acx^2 + (ad+bc)x + bd = (ax+b)(cx+d)$

$x^3 + y^3 = (x+y)(x^2 - xy + y^2)$

$x^3 - y^3 = (x-y)(x^2 + xy + y^2)$

$ax^2 + bx + c = 0$ has roots $x = \frac{-b \pm \sqrt{b^2 - 4ac}}{2a}$

$a^m \cdot a^n = a^{m+n}$ $(a^m)^n = a^{mn}$ $(ab)^n = a^n b^n$ $a^{-n} = 1/a^n$

$a^{p/q} = \sqrt[q]{a^p} = (\sqrt[q]{a})^p$

Important Laws and Formulas

Triangle: Let the angles be *A, B,* and *C* and let the lengths of the opposite sides be *a, b,* and *c.* Then:

$A + B + C = 180^\circ$

Area $= \frac{1}{2}$(length of side)(length of altitude on that side).

Law of sines: $\frac{\sin A}{a} = \frac{\sin B}{b} = \frac{\sin C}{c}$

Law of cosines: $c^2 = a^2 + b^2 - 2ab \cos C$

$b^2 = a^2 + c^2 - 2ac \cos B$

$a^2 = b^2 + c^2 - 2bc \cos A$

Circle: Let r be the radius and θ be the radian measure of the central angle of a sector. Then:

Circumference $(C) = 2\pi r$

Area $(A) = \pi r^2$

Area of sector $= \frac{1}{2} r^2 \theta$

Length of subtended *arc* $= r\theta$

TABLE A-3
Some Geometric Properties

A = plane area. V = volume. C = circumference	Trapezoid: $A = \frac{1}{2}(B + b)h$
S = surface area. p = perimeter	Triangle: $A = \frac{1}{2}bh$
Annulus (ring): $A = \pi(r_2^2 - r_1^2)$	Cube: $V = s^3$, $S = 6s^2$
Circle: $A = \pi r^2$, $D = 2r$, $C = \pi D = 2\pi r$	Rectangular parallelpiped: $V = lwh$, $S = 2(lw + wh + lh)$
Circular sector: $A = \frac{1}{2}r^2\theta$ (θ in radians)	Right circular cone: $V = \frac{1}{3}\pi r^2 h$, $S = \pi r^2 + \pi rl$
Parallelogram: $A = bh$	Right circular cylinder: $V = \pi r^2 h$, $S = 2\pi r^2 + 2\pi rh$
Rectangle: $A = bh$, $p = 2(b + h)$	Sphere: $V = \frac{4}{3}\pi r^3$, $S = 4\pi r^2$
Square: $A = s^2$, $p = 4s$	

TABLE A-4

Table of Conversions

Linear Measure to Metric

1 in. = 2.54 cm
12 in. = 1 ft = 30.48 cm
3 ft = 1 yd = 0.91 m
$5\frac{1}{2}$ yd = 1 rod = 5.03 m
5280 ft = 1 mi = 1.61 km

Linear Measure to English

1 mm = 0.04 in.
1 cm = 0.39 in.
1 m = 39.37 in.
1 km = 0.62 mi

Square Measure to Metric

1 in^2 = 6.45 cm^2
144 in^2 = 1 ft^2 = 0.09 m^2
9 ft^2 = 1 yd^2 = 0.84 m^2
160 rod^2 = 1 acre = 4047.9 m^2
640 acres = 1 mi^2 = 2.59 km^2

Square Measure to English

1 mm^2 = 0.002 in^2
1 cm^2 = 0.16 in^2
1 m^2 = 1549 in^2
1 km^2 = 0.39 mi^2 = 247.10 acres

Cubic Measure to Metric

1 in^3 = 16.39 cm^3
1728 in^3 = 1 ft^3 = 0.03 m^3
27 ft^3 = 1 yd^3 = 0.76 m^3

Cubic Measure to English

1 mm^3 = 0.000061 in^3
1 cm^3 = 0.061 in^3
1 m^3 = 35.32 ft^3
1 km^3 = 0.24 mi^3

Liquid Measure to Metric

1.81 in^3 = 1 fluid oz = 30 mℓ
1 pt = 0.47 ℓ
57.75 in^3 = 1 qt = 0.95 ℓ
231 in^3 = 1 gal = 3.79 ℓ = 0.0038 m^3
1 ft^3 = 7.48 gal = 28.35 ℓ

Liquid Measure to English

1 mℓ = 0.03 fluid oz = 0.061 in^3
1000 cm^3 = 1 ℓ = 61.02 in^3 = 1.06 qt
Note: 1 ft^3 water = 62.5 lb

Weights to Metric

1 oz = 28.35 grams (g)
1 lb = 453.59 g
1 lb = 0.45 kg
1 ton = 0.91 metric ton

Weights to English

1 g = 0.035 oz
1 kg = 2.20 lb
1 metric ton = 1000 kg
= 1.102 tons = 2205 lb

Temperature Formulas

$$C = \frac{5}{9}(F - 32)$$

$$F = \frac{9}{5}C + 32$$

TABLE A-5

Metric-to-English Conversion

Metric Measures of Length

10 millimeters (mm) = 1 centimeter (cm)
10 centimeters = 1 decimeter (dm)
10 decimeters = 1 meter (m)
10 meters = 1 dekameter (dam)
10 dekameters = 1 hectometer (hm)
10 hectometers = 1 kilometer (km)

Conversion Table

1 meter = 39.37 inches = 3.28083 feet = 1.0936 yards
1 centimeter = 0.3937 inch
1 millimeter = 0.03937 inch = $\frac{1}{25}$ inch, approximately
1 kilometer = 0.62137 mile
1 foot = 0.3048 meter
1 inch = 2.54 centimeters = 25.4 millimeters

Table for Converting Millimeters to Inches and Decimals

mm in.	*mm in.*	*mm in.*	*mm in.*
1 = 0.03937	26 = 1.02362	51 = 2.00787	76 = 2.99212
2 = 0.07874	27 = 1.06299	52 = 2.04724	77 = 3.03149
3 = 0.11811	28 = 1.10236	53 = 2.08661	78 = 3.07086
4 = 0.15748	29 = 1.14173	54 = 2.12598	79 = 3.11023
5 = 0.19685	30 = 1.18110	55 = 2.16535	80 = 3.14960
6 = 0.23622	31 = 1.22047	56 = 2.20472	81 = 3.18897
7 = 0.27559	32 = 1.25984	57 = 2.24409	82 = 3.22834
8 = 0.31496	33 = 1.29921	58 = 2.28346	83 = 3.26771
9 = 0.35433	34 = 1.33858	59 = 2.32283	84 = 3.30708
10 = 0.39370	35 = 1.37795	60 = 2.36220	85 = 3.34645
11 = 0.43307	36 = 1.41732	61 = 2.40157	86 = 3.38582
12 = 0.47244	37 = 1.45669	62 = 2.44094	87 = 3.42519
13 = 0.51181	38 = 1.49606	63 = 2.48031	88 = 3.46456
14 = 0.55118	39 = 1.53543	64 = 2.51968	89 = 3.50393
15 = 0.59055	40 = 1.57480	65 = 2.55905	90 = 3.54330
16 = 0.62992	41 = 1.61417	66 = 2.59842	91 = 3.58267
17 = 0.66929	42 = 1.65354	67 = 2.63779	92 = 3.62204
18 = 0.70866	43 = 1.69291	68 = 2.66716	93 = 3.66141
19 = 0.74803	44 = 1.73228	69 = 2.71653	94 = 3.70078
20 = 0.78740	45 = 1.77165	70 = 2.75590	95 = 3.74015
21 = 0.82677	46 = 1.81102	71 = 2.79527	96 = 3.77952
22 = 0.86614	47 = 1.85039	72 = 2.83464	97 = 3.81889
23 = 0.90551	48 = 1.88976	73 = 2.87401	98 = 3.85826
24 = 0.94488	49 = 1.92913	74 = 2.91338	99 = 3.89763
25 = 0.98425	50 = 1.96850	75 = 2.95275	100 = 3.93700

TABLE A-6
Conversion Equivalents

One	Is equal to:
Units of length	
Angstrom unit	10^{-10} meter
Centimeter	0.032 808 4 foot = 0.393 701 inch
Fathom	6 feet
Foot	0.3048 meter
Furlong	$\frac{1}{8}$ mile (40 rods or 1000 chains)[a]
Inch	2.54 centimeters (by law)
Kilometer	0.621 371 miles = 3,280.84 feet
Meter	3.280 84 feet = 1.093 61 yards
Micron	0.001 millimeter = 10,000 angstrom units
Mil	0.001 inch = 25.4 microns
Millimeter	0.039 370 1 inch
Mile	1.609 34 kilometers = 0.868 98 nautical mile
Nautical mile	1.150 78 statute miles = 6,076.115 5 feet
Rod	16.5 feet ($\frac{1}{4}$ of a Gunter's chain or 25 links)[a]
Yard	0.9144 meter
Units of area	
Acre	43,560 sq ft = 4,046.85 sq m = 0.001 562 sq mile
Circular mil	(Defined: area of a circle with a diameter of 1 mil) = 0.7854 sq mil = 5.0671×10^{-6} sq cm
Hectare	10,000 sq m = 2.471 05 acres = 107,637 sq ft
Square centimeter	0.155 00 sq in. = 0.001 076 sq ft = 197,350 circular mils
Square foot	0.092 903 00 sq m = 2.296×10^{-5} acre – 929.0 sq cm = 3.587×10^{-8} sq mile
Square inch	6.4516 sq cm = 1.273×10^{6} circular mils = 7.716×10^{-4} sq yd
Square kilometer	0.3861 sq mile = 1.076×10^{7} sq ft = 247,104 acres
Square meter	10.7639 sq ft = 1.195 99 sq yd
Square mile	2.788×10^{7} sq ft = 2.590 sq km = 640 acres
Square yard	0.836 127 sq m
Square (architects')	100 sq ft

[a]Units in parentheses are obsolete but appear in old deeds.

TABLE A-7
Weights of Materials

	Average weight (lbs)	
Material	*Per cubic foot*	*Per cubic inch*
Aluminum	160	0.0924
Brass	512	0.2960
Brick		
Pressed	150	
Common, hard	125	
Cement		
American, Rosendale	56	
Portland	90	
Clay, loose	63	
Coal, broken, loose		
Anthracite	54	
Bituminous	49	
Concrete	154	
Copper	550	0.3184
Earth		
Common loam	76	
Packed	95	
Gravel		
Dry, loose	90–106	
Well shaken	99–117	
Gold	1206	0.6975
Ice	58.7	
Iron		
Cast	450	0.2600
Wrought	480	0.2778
Lead	710	0.4109
Lime	53	
Masonry		
Well dressed	165	
Dry rubble	138	
Mortar, hardened	103	
Nickel	549	0.3177
Quartz	165	
Sand		
Dry, loose	90–106	
Well shaken	99–117	

TABLE A-7 *(Continued)*

Material	*Average weight (lbs)* Per cubic foot	Per cubic inch
Silver	657	0.3802
Snow		
Freshly fallen	5–12	
Wet and compacted	15–50	
Steel	490	0.2835
Stone		
Gneiss	168	
Granite	170	
Limestone	168	
Marble	168	
Sandstone	151	
Shale	162	
Slate	175	
Tar	62	
Tin	455	0.2632
Water	62.5	
Wood, dry		
Ash	38	
Cherry	42	
Chestnut	41	
Elm	35	
Hemlock	25	
Hickory	53	
Lignum vitae	83	
Mahogany	53	
Maple	49	
Oak		
White	50	
Other kinds	32–45	
Pine		
White	25	
Yellow	34–45	
Spruce	25	
Sycamore	37	
Walnut, black	38	
Zinc	438	0.2528

TABLE A-8

Trigonometric Functions

Degrees	Radians	sin	cos	tan	cot	sec	csc		
0° 00′	.0000	.0000	1.0000	.0000	——	1.000	——	1.5708	**90° 00′**
10	029	029	000	029	343.8	000	343.8	679	50
20	058	058	000	058	171.9	000	171.9	650	40
30	.0087	.0087	1.0000	.0087	114.6	1.000	114.6	1.5621	30
40	116	116	.9999	116	85.94	000	85.95	592	20
50	145	145	999	145	68.75	000	68.76	563	10
1° 00′	.0175	.0175	.9998	.0175	57.29	1.000	57.30	1.5533	**89° 00′**
10	204	204	998	204	49.10	000	49.11	504	50
20	233	233	997	233	42.96	000	42.98	475	40
30	.0262	.0262	.9997	.0262	38.19	1.000	38.20	1.5446	30
40	291	291	996	291	34.37	000	34.38	417	20
50	320	320	995	320	31.24	001	31.26	388	10
2° 00′	.0349	.0349	.9994	.0349	28.64	1.001	28.65	1.5359	**88° 00′**
10	378	378	993	378	26.43	001	26.45	330	50
20	407	407	992	407	24.54	001	24.56	301	40
30	.0436	.0436	.9990	.0437	22.90	1.001	22.93	1.5272	30
40	465	465	989	466	21.47	001	21.49	243	20
50	495	494	988	495	20.21	001	20.23	213	10
3° 00′	.0524	.0523	.9986	.0524	19.08	1.001	19.11	1.5184	**87° 00′**
10	553	552	985	553	18.07	002	18.10	155	50
20	582	581	983	582	17.17	002	17.20	126	40
30	.0611	.0610	.9981	.0612	16.35	1.002	16.38	1.5097	30
40	640	640	980	641	15.60	002	15.64	068	20
50	669	669	978	670	14.92	002	14.96	039	10
4° 00′	.0698	.0698	.9976	.0699	14.30	1.002	14.34	1.5010	**86° 00′**
10	727	727	974	729	13.73	003	13.76	981	50
20	756	756	971	758	13.20	003	13.23	952	40
30	.0785	.0785	.9969	.0787	12.71	1.003	12.75	1.4923	30
40	814	814	967	816	12.25	003	12.29	893	20
50	844	843	964	846	11.83	004	11.87	864	10
5° 00′	.0873	.0872	.9962	.0875	11.43	1.004	11.47	1.4835	**85° 00′**
10	902	901	959	904	11.06	004	11.10	806	50
20	931	929	957	934	10.71	004	10.76	777	40
30	.0960	.0958	.9954	.0963	10.39	1.005	10.43	1.4748	30
40	989	987	951	992	10.08	005	10.13	719	20
50	.1018	.1016	948	.1022	9.788	005	9.839	690	10
6° 00′	.1047	.1045	.9945	.1051	9.514	1.006	9.567	1.4661	**84° 00′**
10	076	074	942	080	9.255	006	9.309	632	50
20	105	103	939	110	9.010	006	9.065	603	40
30	.1134	.1132	.9936	.1139	8.777	1.006	8.834	1.4573	30
40	164	161	932	169	8.556	007	8.614	544	20
50	193	190	929	198	8.345	007	8.405	515	10
7° 00′	.1222	.1219	.9925	.1228	8.144	1.008	8.206	1.4486	**83° 00′**
10	251	248	922	257	7.953	008	8.016	457	50
20	280	276	918	287	7.770	008	7.834	428	40
30	.1309	.1305	.9914	.1317	7.596	1.009	7.661	1.4399	30
40	338	334	911	346	7.429	009	7.496	370	20
50	367	363	907	376	7.269	009	7.337	341	10
8° 00′	.1396	.1392	.9903	.1405	7.115	1.010	7.185	1.4312	**82° 00′**
10	425	421	899	435	6.968	010	7.040	283	50
20	454	449	894	465	6.827	011	6.900	254	40
30	.1484	.1478	.9890	.1495	6.691	1.011	6.765	1.4224	30
40	513	507	886	524	6.561	012	6.636	195	20
50	542	536	881	554	6.435	012	6.512	166	10
9° 00′	.1571	.1564	.9877	.1584	6.314	1.012	6.392	1.4137	**81° 00′**
		cos	sin	cot	tan	csc	sec	Radians	Degrees

Source: *Algebra and Trigonometry: A Pre-Calculus Approach,* 2/e, Max. A. Sobel and Norbert Lerner, Prentice-Hall, Inc., © 1985.

TABLE A-8 *(Continued)*

Degrees	Radians	sin	cos	tan	cot	sec	csc		
9° 00′	.1571	.1564	.9877	.1584	6.314	1.012	6.392	1.4137	**81° 00′**
10	600	593	872	614	197	013	277	108	50
20	629	622	868	644	084	013	166	079	40
30	.1658	.1650	.9863	.1673	5.976	1.014	6.059	1.4050	30
40	687	679	858	703	871	014	5.955	1.4021	20
50	716	708	853	733	769	015	855	992	10
10° 00′	.1745	.1736	.9848	.1763	5.671	1.015	5.759	1.3963	**80° 00′**
10	774	765	843	793	576	016	665	934	50
20	804	794	838	823	485	016	575	904	40
30	.1833	.1822	.9833	.1853	5.396	1.017	5.487	1.3875	30
40	862	851	827	883	309	018	403	846	20
50	891	880	822	914	226	018	320	817	10
11° 00′	.1920	.1908	.9816	.1944	5.145	1.019	5.241	1.3788	**79° 00′**
10	949	937	811	974	066	019	164	759	50
20	978	965	805	.2004	4.989	020	089	730	40
30	.2007	.1994	.9799	.2035	4.915	1.020	5.016	1.3701	30
40	036	.2022	793	065	843	021	4.945	672	20
50	065	051	787	095	773	022	876	643	10
12° 00′	.2094	.2079	.9781	.2126	4.705	1.022	4.810	1.3614	**78° 00′**
10	123	108	775	156	638	023	745	584	50
20	153	136	769	186	574	024	682	555	40
30	.2182	.2164	.9763	.2217	4.511	1.024	4.620	1.3526	30
40	211	193	757	247	449	025	560	497	20
50	240	221	750	278	390	026	502	468	10
13° 00′	.2269	.2250	.9744	.2309	4.331	1 026	4.445	1.3439	**77° 00′**
10	298	278	737	339	275	027	390	410	50
20	327	306	730	370	219	028	336	381	40
30	.2356	.2334	.9724	.2401	4.165	1.028	4.284	1.3352	30
40	385	363	717	432	113	029	232	323	20
50	414	391	710	462	061	030	182	294	10
14° 00′	.2443	.2419	.9703	.2493	4.011	1.031	4.134	1.3265	**76° 00′**
10	473	447	696	524	3.962	031	086	235	50
20	502	476	689	555	914	032	039	206	40
30	.2531	.2504	.9681	.2586	3.867	1.033	3.994	1.3177	30
40	560	532	674	617	821	034	950	148	20
50	589	560	667	648	776	034	906	119	10
15° 00′	.2618	.2588	.9659	.2679	3.732	1.035	3.864	1.3090	**75° 00′**
10	647	616	652	711	689	036	822	061	50
20	676	644	644	742	647	037	782	032	40
30	.2705	.2672	.9636	.2773	3.606	1.038	3.742	1.3003	30
40	734	700	628	805	566	039	703	974	20
50	763	728	621	836	526	039	665	945	10
16° 00′	.2793	.2756	.9613	.2867	3.487	1.040	3.628	1.2915	**74° 00′**
10	822	784	605	899	450	041	592	886	50
20	851	812	596	931	412	042	556	857	40
30	.2880	.2840	.9588	.2962	3.376	1.043	3.521	1.2828	30
40	909	868	580	994	340	044	487	799	20
50	938	896	572	.3026	305	045	453	770	10
17° 00′	.2967	.2924	.9563	.3057	3.271	1.046	3.420	1.2741	**73° 00′**
10	996	952	555	089	237	047	388	712	50
20	.3025	979	546	121	204	048	356	683	40
30	.3054	.3007	.9537	.3153	3.172	1.049	3.326	1.2654	30
40	083	035	528	185	140	049	295	625	20
50	113	062	520	217	108	050	265	595	10
18° 00′	.3142	.3090	.9511	.3249	3.078	1.051	3.236	1.2566	**72° 00′**
		cos	sin	cot	tan	csc	sec	Radians	Degrees

TABLE A-8 *(Continued)*

Degrees	Radians	sin	cos	tan	cot	sec	csc		
18° 00′	.3142	.3090	.9511	.3249	3.078	1.051	3.236	1.2566	**72° 00′**
10	171	118	502	281	047	052	207	537	50
20	200	145	492	314	018	053	179	508	40
30	.3229	.3173	.9483	.3346	2.989	1.054	3.152	1.2479	30
40	258	201	474	378	960	056	124	450	20
50	287	228	465	411	932	057	098	421	10
19° 00′	.3316	.3256	.9455	.3443	2.904	1.058	3.072	1.2392	**71° 00′**
10	345	283	446	476	877	059	046	363	50
20	374	311	436	508	850	060	021	334	40
30	.3403	.3338	.9426	.3541	2.824	1.061	2.996	1.2305	30
40	432	365	417	574	798	062	971	275	20
50	462	393	407	607	773	063	947	246	10
20° 00′	.3491	.3420	.9397	.3640	2.747	1.064	2.924	1.2217	**70° 00′**
10	520	449	387	673	723	065	901	188	50
20	549	475	377	706	699	066	878	159	40
30	.3578	.3502	.9367	.3739	2.675	1.068	2.855	1.2130	30
40	607	529	356	772	651	069	833	101	20
50	636	557	346	805	628	070	812	072	10
21°00′	.3665	.3584	.9336	.3839	2.605	1.071	2.790	1.2043	**69° 00′**
10	694	611	325	872	583	072	769	1.2014	50
20	723	638	315	906	560	074	749	985	40
30	.3752	.3665	.9304	.3939	2.539	1.075	2.729	1.1956	30
40	782	692	293	973	517	076	709	926	20
50	811	719	283	.4006	496	077	689	897	10
22° 00′	.3840	.3746	.9272	.4040	2.475	1.079	2.669	1.1868	**68° 00′**
10	869	773	261	074	455	080	650	839	50
20	898	800	250	108	434	081	632	810	40
30	.3927	.3827	.9239	.4142	2.414	1.082	2.613	1.1781	30
40	956	854	228	176	394	084	595	752	20
50	985	881	216	210	375	085	577	723	10
23° 00′	.4014	.3907	.9205	.4245	2.356	1.086	2.559	1.1694	**67° 00′**
10	043	934	194	279	337	088	542	665	50
20	072	961	182	314	318	089	525	636	40
30	.4102	.3987	.9171	.4348	2.300	1.090	2.508	1.1606	30
40	131	.4014	159	383	282	092	491	577	20
50	160	041	147	417	264	093	475	548	10
24° 00′	.4189	.4067	.9135	.4452	2.246	1.095	2.459	1.1519	**66° 00′**
10	218	094	124	487	229	096	443	490	50
20	247	120	112	522	211	097	427	461	40
30	.4276	.4147	.9100	.4557	2.194	1.099	2.411	1.1432	30
40	305	173	088	592	177	100	396	403	20
50	334	200	075	628	161	102	381	374	10
25° 00′	.4363	.4226	.9063	.4663	2.145	1.103	2.366	1.1345	**65° 00′**
10	392	253	051	699	128	105	352	316	50
20	422	279	038	734	112	106	337	286	40
30	.4451	.4305	.9026	.4770	2.097	1.108	2.323	1.1257	30
40	480	331	013	806	081	109	309	228	20
50	509	358	001	841	066	111	295	199	10
26° 00′	.4538	.4384	.8988	.4877	2.050	1.113	2.281	1.1170	**64° 00′**
10	567	410	975	913	035	114	268	141	50
20	596	436	962	950	020	116	254	112	40
30	.4625	.4462	.8949	.4986	2.006	1.117	2.241	1.1083	30
40	654	488	936	.5022	1.991	119	228	054	20
50	683	514	923	059	977	121	215	1.1025	10
27° 00′	.4712	.4540	.8910	.5095	1.963	1.122	2.203	1.0996	**63° 00′**
		cos	sin	cot	tan	csc	sec	Radians	Degrees

TABLE A-8 *(Continued)*

Degrees	Radians	sin	cos	tan	cot	sec	csc		
27° 00′	.4712	.4540	.8910	.5095	1.963	1.122	2.203	1.0996	**63° 00′**
10	741	566	897	132	949	124	190	966	50
20	771	592	884	169	935	126	178	937	40
30	.4800	.4617	.8870	.5206	1.921	1.127	2.166	1.0908	30
40	829	643	857	243	907	129	154	879	20
50	858	669	843	280	894	131	142	850	10
28° 00′	.4887	.4695	.8829	.5317	1.881	1.133	2.130	1.0821	**62° 00′**
10	916	720	816	354	868	134	118	792	50
20	945	746	802	392	855	136	107	763	40
30	.4974	.4772	.8788	.5430	1.842	1.138	2.096	1.0734	30
40	.5003	797	774	467	829	140	085	705	20
50	032	823	760	505	816	142	074	676	10
29° 00′	.5061	.4848	.8746	.5543	1.804	1.143	2.063	1.0647	**61° 00′**
10	091	874	732	581	792	145	052	617	50
20	120	899	718	619	780	147	041	588	40
30	.5149	.4924	.8704	.5658	1.767	1.149	2.031	1.0559	30
40	178	950	689	696	756	151	020	530	20
50	207	975	675	735	744	153	010	501	10
30° 00′	.5236	.5000	.8660	.5774	1.732	1.155	2.000	1.0472	**60° 00′**
10	265	025	646	812	720	157	1.990	443	50
20	294	050	631	851	709	159	980	414	40
30	.5323	.5075	.8616	.5890	1.698	1.161	1.970	1.0385	30
40	352	100	601	930	686	163	961	356	20
50	381	125	587	969	675	165	951	327	10
31° 00′	.5411	.5150	.8572	.6009	1.664	1.167	1.942	1.0297	**59° 00′**
10	440	175	557	048	653	169	932	268	50
20	469	200	542	088	643	171	923	239	40
30	.5498	.5225	.8526	.6128	1.632	1.173	1.914	1.0210	30
40	527	250	511	168	621	175	905	181	20
50	556	275	496	208	611	177	896	152	10
32° 00′	.5585	.5299	.8480	.6249	1.600	1.179	1.887	1.0123	**58° 00′**
10	614	324	465	289	590	181	878	094	50
20	643	348	450	330	580	184	870	065	40
30	.5672	.5373	.8434	.6371	1.570	1.186	1.861	1.0036	30
40	701	398	418	412	560	188	853	1.0007	20
50	730	422	403	453	550	190	844	977	10
33° 00′	.5760	.5446	.8387	.6494	1.540	1.192	1.836	.9948	**57° 00′**
10	789	471	371	536	530	195	828	919	50
20	818	495	355	577	520	197	820	890	40
30	.5847	.5519	.8339	.6619	1.511	1.199	1.812	.9861	30
40	876	544	323	661	501	202	804	832	20
50	905	568	307	703	1.492	204	796	803	10
34° 00′	.5934	.5592	.8290	.6745	1.483	1.206	1.788	.9774	**56° 00′**
10	963	616	274	787	473	209	731	745	50
20	992	640	258	830	464	211	773	716	40
30	.6021	.5664	.8241	.6873	1.455	1.213	1.766	.9687	30
40	050	688	225	916	446	216	758	657	20
50	080	712	208	959	437	218	751	628	10
35° 00′	.6109	.5736	.8192	.7002	1.428	1.221	1.743	.9599	**55° 00′**
10	138	760	175	046	419	223	736	570	50
20	167	783	158	089	411	226	729	541	40
30	.6196	.5807	.8141	.7133	1.402	1.228	1.722	.9512	30
40	225	831	124	177	393	231	715	483	20
50	254	854	107	221	385	233	708	454	10
36° 00′	.6283	.5878	.8090	.7265	1.376	1.236	1.701	.9425	**54° 00′**
		cos	sin	cot	tan	csc	sec	Radians	Degrees

TABLE A-8 *(Continued)*

Degrees	Radians	sin	cos	tan	cot	sec	csc		
36° 00′	.6283	.5878	.8090	.7265	1.376	1.236	1.701	.9425	**54° 00′**
10	312	901	073	310	368	239	695	396	50
20	341	925	056	355	360	241	688	367	40
30	.6370	.5948	.8039	.7400	1.351	1.244	1.681	.9338	30
40	400	972	021	445	343	247	675	308	20
50	429	995	004	490	335	249	668	279	10
37° 00′	.6458	.6018	.7986	.7536	1.327	1.252	1.662	.9250	**53° 00′**
10	487	041	969	581	319	255	655	221	50
20	516	065	951	627	311	258	649	192	40
30	.6545	.6088	.7934	.7673	1.303	1.260	1.643	.9163	30
40	574	111	916	720	295	263	636	134	20
50	603	134	898	766	288	266	630	105	10
38° 00′	.5632	.6157	.7880	.7813	1.280	1.269	1.624	.9076	**52° 00′**
10	661	180	862	860	272	272	618	047	50
20	690	202	844	907	265	275	612	.9018	40
30	.6720	.6225	.7826	.7954	1.257	1.278	1.606	.8988	30
40	749	248	808	.8002	250	281	601	959	20
50	778	271	790	050	242	284	595	930	10
39° 00′	.6807	.6293	.7771	.8098	1.235	1.287	1.589	.8901	**51° 00′**
10	836	316	753	146	228	290	583	872	50
20	865	338	735	195	220	293	578	843	40
30	.6894	.6361	.7716	.8243	1.213	1.296	1.572	.8814	30
40	923	383	698	292	206	299	567	785	20
50	952	406	679	342	199	302	561	756	10
40° 00′	.6981	.6428	.7660	.8391	1.192	1.305	1.556	.8727	**50° 00′**
10	.7010	450	642	441	185	309	550	698	50
20	039	472	623	491	178	312	545	668	40
30	.7069	.6494	.7604	.8541	1.171	1.315	1.540	.8639	30
40	098	517	585	591	164	318	535	610	20
50	127	539	566	642	157	322	529	581	10
41° 00′	.7156	.6561	.7547	.8693	1.150	1.325	1.524	.8552	**49° 00′**
10	185	583	528	744	144	328	519	523	50
20	214	604	509	796	137	332	514	494	40
30	.7243	.6626	.7490	.8847	1.130	1.335	1.509	.8465	30
40	272	648	470	899	124	339	504	436	20
50	301	670	451	952	117	342	499	407	10
42° 00′	.7330	.6691	.7431	.9004	1.111	1.346	1.494	.8378	**48° 00′**
10	359	713	412	057	104	349	490	348	50
20	389	734	392	110	098	353	485	319	40
30	.7418	.6756	.7373	.9163	1.091	1.356	1.480	.8290	30
40	447	777	353	217	085	360	476	261	20
50	476	799	333	271	079	364	471	232	10
43° 00′	.7505	.6820	.7314	.9325	1.072	1.367	1.466	.8203	**47° 00′**
10	534	841	294	380	066	371	462	174	50
20	563	862	274	435	060	375	457	145	40
30	.7592	.6884	.7254	.9490	1.054	1.379	1.453	.8116	30
40	621	905	234	545	048	382	448	087	20
50	650	926	214	601	042	386	444	058	10
44° 00′	.7679	.6947	.7193	.9657	1.036	1.390	1.440	.8029	**46° 00′**
10	709	967	173	713	030	394	435	999	50
20	738	988	153	770	024	398	431	970	40
30	.7767	.7009	.7133	.9827	1.018	1.402	1.427	.7941	30
40	796	030	112	884	012	406	423	912	20
50	825	050	092	942	006	410	418	883	10
45° 00′	.7854	.7071	.7071	1.000	1.000	1.414	1.414	.7854	**45° 00′**
		cos	sin	cot	tan	csc	sec	Radians	Degrees

Appendix

B

PRETESTS AND POSTTESTS

PRETEST

Chapter 1

Express the place value of the digit underlined in each number.

1. $34\underline{2},153$ 2. $1,\underline{0}00$

Expand each whole number.

3. 6,942 4. 5

Write each whole number as a word statement.

5. 864 6. 53,084

Complete each problem.

7. $\begin{array}{r} 304 \\ +\ \underline{\ \ 69} \end{array}$ 8. $\begin{array}{r} 4,985 \\ 69,333 \\ +\ \underline{\ 5,009} \end{array}$ 9. $\begin{array}{r} 12,649 \\ -\ \underline{\ \ \ 385} \end{array}$ 10. $\begin{array}{r} 698 \\ -\ \underline{179} \end{array}$

11. $\begin{array}{r} 837 \\ \times\ \underline{\ \ \ 5} \end{array}$ 12. $\begin{array}{r} 69,483 \\ \times\ \underline{\ \ \ \ \ \ 8} \end{array}$ 13. $\begin{array}{r} 1,984 \\ \times\ \underline{\ \ 444} \end{array}$ 14. $\begin{array}{r} 795 \\ \times\ \underline{795} \end{array}$

15. $69\overline{)984}$ 16. $22\overline{)689}$

POSTTEST

Chapter 1

Express the place value of the digit underlined in each number.

1. $3{,}\underline{0}98$

2. $\underline{9}$

Expand each whole number.

3. 896,482

4. 7,010

Write each whole number as a word statement.

5. 3,020

6. 645,123

Complete each problem.

7. $\begin{array}{r} 3{,}987 \\ +\ \underline{\ \ 123} \end{array}$

8. $\begin{array}{r} 998 \\ 69 \\ +\ \underline{2{,}564} \end{array}$

9. $\begin{array}{r} 695 \\ -\ \underline{\ 35} \end{array}$

10. $\begin{array}{r} 83 \\ -\ \underline{69} \end{array}$

11. $\begin{array}{r} 246 \\ \times\ \underline{\ \ 8} \end{array}$

12. $\begin{array}{r} 932 \\ \times\ \underline{\ \ 4} \end{array}$

13. $\begin{array}{r} 894 \\ \times\ \underline{\ 29} \end{array}$

14. $\begin{array}{r} 685 \\ \times\ \underline{\ 32} \end{array}$

15. $29\overline{)6{,}945}$

16. $8\overline{)39}$

PRETEST

Chapter 2

Express each fraction as an equivalent.

1. 5/6 = ?/30
2. 3/9 = ?/27

Express each fraction in its simplest form.

3. 8/56
4. 9/27

Complete each addition problem. Show all work. Express each answer in the simplest terms.

5. 6/8 + 1/8
6. 3/8 + 9/32

Complete each subtraction problem. Show all intermediate steps. Express each answer in the simplest terms.

7. 7/8 − 3/4
8. 7/18 − 3/24

Complete each multiplication problem. Express each answer in the simplest terms.

9. 7 × 1/7
10. 3 × 5/6
11. 3/4 × 6/7
12. 5/11 × 1/2

Complete each division problem. Express each answer in the simplest terms.

13. 2/4 ÷ 3
14. 9/64 ÷ 2
15. 3 ÷ 1/5
16. 6 ÷ 1/3
17. 3/4 ÷ 1/2
18. 1/2 ÷ 5/7

POSTTEST

Chapter 2

Express each fraction as an equivalent.

1. 8/9 = ?/18
2. 3/8 = ?/64

Express each fraction in its simplest form.

3. 11/22
4. 62/30

Complete each addition problem. Show all work. Express each answer in the simplest terms.

5. 3/4 + 1/4
6. 3/8 + 3/16

Complete each subtraction problem. Show all intermediate steps. Express each answer in the simplest terms.

7. 42/133 − 25/133
8. 4/9 − 1/5

Complete each multiplication problem. Express each answer in the simplest terms.

9. 6 × 2/3
10. 11 × 3/16
11. 6/11 × 1/2
12. 7/32 × 9/64

Complete each division problem. Express each answer in the simplest terms.

13. 2/3 ÷ 6
14. 7/8 ÷ 12
15. 2 ÷ 3/4
16. 12 ÷ 2/5
17. 1/2 ÷ 1/2
18. 1/4 ÷ 2/3

PRETEST

Chapter 3

Complete each addition problem. Express each answer in the simplest terms.

1. 2 3/4 + 4 1/8

2. 2 13/45 + 3 1/5

Complete each subtraction problem. Express each answer in the simplest terms.

3. 3 1/5 − 2 1/4

4. 42 1/6 − 3 1/7

Complete each multiplication problem. Express each answer in the simplest terms.

5. 4 1/4 × 16 1/2

6. 7 3/4 × 5/6

Complete each division problem. Express each answer in the simplest terms.

7. 16 3/4 ÷ 9 1/16

8. 7 1/2 ÷ 3 1/3

POSTTEST

Chapter 3

Complete each addition problem. Express each answer in the simplest terms.

1. 6 1/2 + 9 1/5

2. 3 1/3 + 4 1/4

Complete each subtraction problem. Express each answer in the simplest terms.

3. 5 1/10 − 3 1/5

4. 145 1/2 − 66 1/3

Complete each multiplication problem. Express each answer in the simplest terms.

5. 5 6/10 × 6 8/9

6. 4 1/4 × 5 1/5

Complete each division problem. Express each answer in the simplest terms.

7. 10 1/3 ÷ 3 1/2

8. 9 1/6 ÷ 4 1/2

PRETEST

Chapter 4

Write out the word interpretation for each decimal fraction or mixed decimal.

1. 0.9

2. 6.89

Convert each common fraction to a decimal fraction and round-off each answer to two places.

3. 3/5

4. 10 2/3

Convert each decimal fraction to a common fraction and reduce each answer to the simplest terms.

5. 6.15

6. 0.7

Complete each addition problem.

7. $\begin{array}{r} 12.93 \\ +\ \underline{14.31} \end{array}$

8. $\begin{array}{r} 52.943 \\ 69.670 \\ +\ \underline{\ \ 1.864} \end{array}$

Complete each subtraction problem.

9. 12.06 − 7.98

10. 1,000.05 − 99.99

Complete each multiplication problem. Round-off each answer to three places.

11. 15.32 × 0.777

12. 0.18 × 0.11

Complete each division problem. Round-off each answer to two places.

13. 0.65 ÷ 0.36

14. 14.617 ÷ 0.13

POSTTEST

Chapter 4

Write out the word interpretations for each mixed decimal.

1. 2.35

2. 25.375

Convert each common fraction to a decimal fraction and round-off each answer to two places.

3. 5 3/4

4. 110 1/8

Convert each decimal fraction to a common fraction and express each answer in the simplest terms.

5. 0.75

6. 20.08

Complete each addition problem.

7.
$$\begin{array}{r} 12.34 \\ 56.78 \\ +\ \underline{91.01} \end{array}$$

8.
$$\begin{array}{r} 12.13 \\ 14.15 \\ +\ \underline{16.00} \end{array}$$

Complete each subtraction problem.

9. 339.8 − 12.3

10. 29.1 − 4.432

Complete each multiplication problem. Round-off each answer to three places.

11. 15.001 × 0.81

12. 1.34 × 66.66

Complete each division problem. Round-off each answer to two places.

13. 10.69 ÷ 0.36

14. 136.62 ÷ 62

PRETEST

Chapter 5

Raise each number to the power indicated by the exponent.

1. 4^3

2. 10^4

Take the indicated roots in each problem. Use an electronic calculator with a root function.

3. $\sqrt{1}$

4. $\sqrt{20.9}$

Express each fraction as a percent.

5. 0.0301

6. 11/12

Convert each percent expression to a decimal.

7. 11%

8. 3 2/7%

Write each percent expression as a common fraction.

9. 20.5%

10. 3.9%

Solve each percentage problem.

11. 12% of 243

12. 0.31% of 100

POSTTEST

Chapter 5

Raise each number to the power indicated by the exponent.

1. 11^2

2. 9.83^3

Take the indicated root in each problem. Use an electronic calculator with a root function.

3. $\sqrt{13}$

4. $\sqrt[3]{20.8}$

Express each fraction as a percent.

5. 0.91

6. 3 12/17

Convert each percent expression to a decimal.

7. 28%

8. 99.3%

Write each percent expression as a common fraction.

9. 2.33%

10. 17.05%

Solve each percentage problem.

11. 16% of 24

12. 1.61% of 11.2

Appendix C

ANSWERS

PRETEST ANSWERS

Chapter 1

1. Thousands 3. $(6 \times 1{,}000) + (9 \times 100) + (4 \times 10) + (2 \times 1)$
5. Eight hundred sixty-four 7. 373 9. 12,264 11. 4,185 13. 880,896
15. 14 R18

POSTTEST ANSWERS

Chapter 1

1. Hundreds
3. $(8 \times 100{,}000) + (9 \times 10{,}000) + (6 \times 1{,}000) + (4 \times 100) + (8 \times 10) + (2 \times 1)$
5. Three thousand twenty 7. 4,110 9. 660 11. 1,968 13. 25,926
15. 239 R14

PRETEST ANSWERS

Chapter 2

1. 25/30 3. 1/7 5. 7/8 7. 1/8 9. 1 11. 18/28 = 9/14 13. 2/12 = 1/6
15. 15 17. 6/4 = 1 1/2

POSTTEST ANSWERS

Chapter 2

1. 16/18   3. 1/2   5. 1   7. 17/133   9. 4   11. 6/22   13. 1/9
15. 8/3 = 2 2/8 = 2 1/4   17. 1

PRETEST ANSWERS

Chapter 3

1. 6 7/8   3. 19/20   5. 70 1/8   7. 1 123/145

POSTTEST ANSWERS

Chapter 3

1. 15 7/10   3. 1 9/10   5. 38 26/45   7. 2 20/21

PRETEST ANSWERS

Chapter 4

1. Nine-tenths   3. 0.6   5. 6 15/100 = 6 3/20   7. 27.24   9. 4.08   11. 11.90
13. 1.81

POSTTEST ANSWERS

Chapter 4

1. Two and thirty-five hundredths   3. 5.75   5. 3/4   7. 160.13   9. 327.5
11. 12.151   13. 29.69

PRETEST ANSWERS

Chapter 5

1. 64   3. 3.317   5. 3.01%   7. 0.11   9. 0.205 = 205/1,000   11. 29.16

POSTTEST ANSWERS

Chapter 5

1. 121   3. 3.61   5. 91%   7. 0.28   9. 2 1/3   11. 3.84

ANSWERS TO PRACTICE PROBLEMS

Practice Problem Set 1-1

1. Tens 3. Hundreds 5. Thousands 7. Units 9. Hundred-thousands

Practice Problem Set 1-2

1. $(1 \times 100) + (7 \times 10) + (2 \times 1)$ 3. $(1 \times 10) + (6 \times 1)$ 5. $(9 \times 10) + (9 \times 1)$
7. $(7 \times 1{,}000) + (5 \times 100) + (2 \times 10) + (4 \times 1)$ 9. $(6 \times 10) + (2 \times 1)$

Practice-Problem Set 1-3

1. Seven hundred nineteen 3. Ten thousand, four hundred nineteen
5. Seven million, three hundred twenty-six thousand, four hundred twenty-four
7. Eight thousand, eight hundred eighty-eight
9. One hundred twelve thousand, five hundred thirty-seven

Practice Problem Set 1-4

1. 91 3. 143 5. 848 7. 1,406 9. 34,953

Practice Problem Set 1-5

1. 45 3. 220 5. 375 7. 8,717 9. 658

Practice Problem Set 1-6

1. 133 3. 2,672 5. 41,058 7. 25,928 9. 111, 126

Practice Problem Set 1-7

1. 1,875 3. 32,994 5. 816,390 7. 917,217 9. 5,517,936

Practice Problem Set 1-8

1. 6 R3 3. 29 R1 5. 214 R1 7. 27 R14 9. 16 R54

Practice Problem Set 2-1

1. 6/8 3. 10/16 5. 18/32 7. 14/64 9. 10/128

Practice Problem Set 2-2

1. 1/2 3. 2/3 5. 3/4 7. 1/2 9. 2/3

Practice Problem Set 2-3

1. 3/3 = 1 3. 8/8 = 1 5. 5/4 = 1 1/4 7. 26/36 9. 37/72

Practice Problem Set 2-4

1. 2/4 = 1/2 3. 8/64 = 1/8 5. 1/8 7. 14/36 = 7/18 9. 3/32

Practice Problem Set 2-5

1. 3 1/3 3. 9/4 = 2 1/4 5. 12/12 = 1 7. 14/4 = 3 1/2 9. 55/8 = 6 7/8

Practice Problem Set 2-6

1. 1/16 3. 7/32 5. 36/50 = 18/25 7. 5/24 9. 285/2112

Practice Problem Set 2-7

1. 2/15 3. 1/96 5. 11/30 7. 1/6 9. 7/108

Practice Problem Set 2-8

1. 15/2 = 7 1/2 3. 96/9 = 10 2/3 5. 30/11 = 2 8/11 7. 6/1 = 6
9. 108/7 = 15 3/7

Practice Problem Set 2-9

1. 4/2 = 2 3. 10/6 = 1 2/3 5. 88/84 = 1 1/21 7. 24/18 = 1 1/3 9. 7/20

Practice Problem Set 3-1

1. 4 3/3 = 5 3. 9 7/8 5. 82 19/32 7. 13 23/50 9. 7 16/45

Practice Problem Set 3-2

1. 1/3 3. 1 5/8 5. 52 13/32 7. 4 41/50 9. 1 2/9

Practice Problem Set 3-3

1. 3 9/16 3. 40 7/64 5. 32 14/15 7. 12 33/40 9. 258 1/2

Practice Problem Set 3-4

1. 1 7/12 3. 2 15/68 5. 1 17/78 7. 2 567/864 9. 4 3/5

Practice Problem Set 4-1

1. One-tenth 3. Seven-hundredths
5. One thousand four hundred fifty-six ten-thousandths
7. Six and eight hundred ninety-two thousandths
9. Three hundred twenty-five and four thousand one hundred ten ten-thousandths

Practice Problem Set 4-2

1. 0.75 3. 0.56 5. 0.03 7. 10.06 9. 18.13

Practice Problem Set 4-3

1. 73/100 3. 55/100 5. 1/25 7. 11 2/25 9. 19 3/20

Practice Problem Set 4-4

1. 66.038 3. 179.011 5. 1,953.97 7. 204.24 9. 1,873.193

Practice Problem Set 4-5

1. 2.270 3. 0.790 5. 56.171 7. 726.024 9. 1,232.943

Practice Problem Set 4-6

1. 546.979 3. 311.874 5. 22,861.608 7. 0.262 9. 0.180

Practice Problem Set 4-7

1. 2.43 3. 1.21 5. 64.88 7. 2.50 9. 1.06

Practice Problem Set 5-1

1. 125 3. 100 5. 2,401 7. 65.12 9. 1,036.43

Practice Problem Set 5-2

1. 3.46 3. 9.30 5. 37.63 7. 3.05 9. 8.74

Practice Problem Set 5-3

1. 2.15% 3. 16% 5. 89% 7. 89% 9. 1,285.7%

Practice Problem Set 5-4

1. 0.68 3. 9.51 5. 14.18 7. 0.2775 9. 0.01375

Practice Problem Set 5-5

1. 39/200 3. 221/2000 5. 37/40 7. 57/5000 9. 171/2000

Practice Problem Set 5-6

1. 7.9 3. 9.75 5. 39.52 7. 49.29 9. 0.0026415

Practice Problem Set 6-1

1. $X + 5$ 3. $35 - B$ 5. $\frac{Y^2}{3 \times 5}$ 7. 5 9. 39

Practice Problem Set 6-2

1. 5 3. 3.4 5. 66.8 7. −16 9. 28 11. 5 13. −25 15. 13.46
17. 132 19. −0.77 21. −10.4 23. −3.57 25. 10.60 27. −27
29. 310.46

Practice Problem Set 6-3

1. $17ab$ 3. $25x^2$ 5. $5x + 18xy + 18x^2y$ 7. $7a^2b$ 9. $13.67x^3y^4$ 11. $5xa$
13. $-12abx$ 15. $22.5x^3 - 5x^2y + 40.5xy - 9y^2$ 17. $3x$ 19. $4c^2d^4 + 9ab$
21. $4a^2 + 4ab + 6^2$ 23. $1000a^3$ 25. $4a^4b^2$

Practice Problem Set 7-1

1. 20 3. 37.6 5. 3.753 7. 161.51 9. 15

Practice Problem Set 7-2

1. 6 3. 16.7 5. 4.318 7. 58 9. 15.97

Practice Problem Set 7-3

1. 250 3. 21.84 5. 105.34 7. 51.53 9. 100.89

Practice Problem Set 7-4

1. 2 3. 3.31 5. 19.03 7. 29.98 9. 166.67

Practice Problem Set 7-5

1. 3 3. 3.95 5. 8.64 7. 10.24 9. 58.06

Practice Problem Set 7-6

1. $a = 5.13$ 3. $x = 12.4$ 5. $d = 5.18$ 7. $x = 11.92$ 9. $p = 7{,}157.16$

Practice Problem Set 8-1

1. Right 3. Acute 5. Acute 7. 90° 9. 77°

Practice Problem Set 8-2

1. 3.25° 3. 33.57° 5. 67.92° 7. 59.17° 9. 45.75°

Practice Problem Set 8-3

1. 9°10′12″ 3. 24°28′12″ 5. 49°28′48″ 7. 113°10′12″ 9. 211°34′48″

Practice Problem Set 8-4

1. 28°29′ 3. 60°11′40″ 5. 271°29′31″ 7. 60°18′22″ 9. 22°07′55″

Practice Problem Set 8-5

1. 76°32′ 3. 208°48′36″ 5. 522°06′10″ 7. 49°39′ 9. 16°10′18″

Practice Problem Set 9-1

1. Scalene 3. Isosceles 5. Equilateral 7. 40° 9. 104° 11. 7.21
13. 11.40 15. 12

Practice Problem Set 9-2

1. 60° 3. 120° 5. 210° 7. 75° 9. 150°

Practice Problem Set 10-1

1. 31.42 3. 23.94 5. 24.57 7. 5.41 9. 52.04

Practice Problem Set 10-2

1. 8.46 3. 5.97 5. 16.17 7. 0.89 9. 13.36

Practice Problem Set 11-1

1. 10 3. 14.21 5. 19.14 7. 13.27 9. 5.86

Practice Problem Set 11-2

1. 0.52 3. 3.31 5. 1.11 7. 0.06 9. 20.79

Practice Problem Set 11-3

1. $B = 78$, $b = 92$, $c = 94.08$ 3. $A = 23$, $a = 27.82$, $b = 65.54$
5. $A = 41$, $b = 4.00$, $c = 5.30$ 7. $a = 3.75$, $B = 76$, $c = 15.49$
9. $A = 35°57'02''$, $B = 54°02'58''$, $a = 12.33$

Practice Problem Set 12-1

1. $A = 66$, $b = 31.48$, $c = 26.83$
3. $B = 51$, $a = 12.39$, $c = 10.92$
5. $C = 75$, $b = 2.26$, $c = 5.84$
7. $B = 58°51'18''$, $C = 88°08'42''$, $c = 12.85$
9. $A = 63°29'25''$, $B = 51°30'35''$, $a = 21.72$

Practice Problem Set 12-2

1. $A = 95°04'01''$, $B = 46°55'59''$, $c = 9.27$
3. $B = 60°45'28''$, $C = 44°14'32''$, $a = 22.14$
5. $B = 78°23'55''$, $C = 58°36'05''$, $a = 31.33$
7. $A = 41°04'31''$, $B = 47°57'59''$, $C = 90°57'30''$
9. $A = 36°12'04''$, $B = 40°01'29''$, $C = 103°46'27''$

Practice Problem Set 13-1

1. 90 square inches 3. 310.50 square inches 5. 28.43 square inches
7. 335.84 square inches 9. 79.14 square inches

Practice Problem Set 13-2

1. 14.52 square inches 3. 76.11 square inches 5. 40.05 square inches
7. 4,799.56 square inches 9. 1,152.66 square inches

ANSWERS TO REVIEW PROBLEMS

Chapter 1

1. Thousands 3. Ten-thousands 5. (2 × 1,000) + (5 × 100) + (4 × 10) + (3 × 1)
7. Six million, nine hundred five thousand, nine hundred 9. 9,135 11. 5,903
13. 1,112 15. 81,918 17. 2,125 19. 16 R10

Chapter 2

1. 49/63 3. 49/56 5. 3/4 7. 6/4 = 1 1/2 9. 5/18 11. 16/3 = 5 1/3
13. 1/3 15. 1/6 17. 16 19. 5/6

Chapter 3

1. 13 3. 10 13/14 5. 7 1/2 7. 2 5/22 9. 8 5/24 11. 68 1/16 13. 28
15. 20 40/49 17. 7 13/15 19. 7 8/9

Chapter 4

1. One hundred one and nine-tenths **3.** One and two-hundredths **5.** 11.67
7. 9 8/100 = 9 2/25 **9.** 4/100 = 1/25 **11.** 55.87 **13.** 27.38 **15.** 1,008.24
17. 161.7 **19.** 2.37

Chapter 5

1. 90.44 **3.** 7.55 **5.** 5.62% **7.** 518.75% **9.** 0.75 **11.** 0.6975
13. 0.09438 **15.** 21 5/100 **17.** 1 16/100 **19.** 28.78

Chapter 6

1. $5X^2$ **3.** 10.955 **5.** 5, 3, 1.76 **7.** 13 **9.** 125 **11.** −70 **13.** −12.40
15. 2,746.81 **17.** $35.9x^3y^2$ **19.** $12y^2x + 9x^2$

Chapter 7

1. $24.17 = X$ **3.** $50.18 = C$ **5.** $6.6414 = X$ **7.** $192.06 = X$ **9.** $50.54 = C$
11. $5.66 = R$ **13.** $4.18 = X$ **15.** $6.14 = a$ **17.** $6.64 = b$ **19.** $10.75 = y$

Chapter 8

1. 4.27° **3.** 36.80° **5.** 69.94° **7.** 69.19° **9.** 55.93° **11.** 16°46′48″
13. 36°43′12″ **15.** 351°44′24″ **17.** 37°14′ **19.** 11°57′59″

Chapter 9

1. A simple, three-sided polygon composed of straight lines and angles **3.** Hypotenuse
5. A triangle with three equal sides **7.** A triangle with three unequal sides
9. The angle opposite a given side **11.** Congruency = exact image **13.** 152°
15. 46.83° **17.** 287° **19.** 89.73°

Chapter 10

1. 37.70 **3.** 17.48 **5.** 7.82 **7.** 6.06 **9.** 52.55 **11.** 13.07 **13.** 10.05
15. 19.87 **17.** 1.66 **19.** 16.64

Chapter 11

1. $c = 11.40$ **3.** $c = 15.62$ **5.** $c = 21.97$ **7.** $a = 13.86$ **9.** $b = 6.37$
11. $B = 77$, $b = 89.10$, $c = 91.44$ **13.** $A = 26$, $a = 31.77$, $b = 65.14$
15. $A = 31$, $b = 13.90$, $c = 16.19$ **17.** $B = 74$, $a = 4.97$, $c = 18.03$
19. $A = 45.27$, $B = 44.73$, $a = 19.18$

Chapter 12

1. $A = 62$, $b = 35.18$, $c = 30.34$ **3.** $B = 43$, $a = 20.15$, $c = 18.16$
5. $C = 78$, $b = 2.70$, $c = 11.74$ **7.** $B = 92$, $C = 44$
9. $A = 63.49$, $B = 51.51$, $a = 21.72$ **11.** $A = 88.75$, $B = 51.25$, $c = 10.93$
13. $B = 56.80$, $C = 44.20$, $a = 28.16$ **15.** $B = 77.50$, $C = 53.50$, $a = 39.43$
17. $A = 46.46$, $B = 52.26$, $C = 81.28$ **19.** $A = 35.62$, $B = 39.44$, $C = 104.94$

Chapter 13

1. 104 **3.** 348 **5.** 272.83 **7.** 291.78 **9.** 137.17 **11.** 18.51 **13.** 64.24
15. 2,912.52 **17.** 3,584.73 **19.** 196.58

INDEX